Parallel Processing in Digital Control

D. Fabian Garcia Nocetti
Peter J. Fleming

Parallel Processing in Digital Control

With 79 Figures

Springer-Verlag
London Berlin Heidelberg New York
Paris Tokyo Hong Kong
Barcelona Budapest

D. Fabian Garcia Nocetti, BSc, PhD
School of Electronic Engineering Science, University of Wales,
Dean Street, Bangor, Gwynedd LL57 1UT, UK

Peter J. Fleming, BSc, PhD, MIEE CEng, MBSC, MInstMC
Automatic Control and Systems Engineering, University of
Sheffield, P.O. Box 600, Mappin Street, Sheffield S1 4DU, UK

Cover illustration: A transputer network

ISBN 3-540-19728-1 Springer-Verlag Berlin Heidelberg New York
ISBN 0-387-19728-1 Springer-Verlag New York Berlin Heidelberg

British Library Cataloguing in Publication Data
A catalogue record for this book is available from the British Library

Library of Congress Cataloging-in-Publication Data
A catalog record for this book is available from the Library of Congress

Apart from any fair dealing for the purposes of research or private study, or criticism or
review, as permitted under the Copyright, Designs and Patents Act 1988, this publication
may only be reproduced, stored or transmitted, in any form or by any means, with the
prior permission in writing of the publishers, or in the case of reprographic reproduction in
accordance with the terms of licences issued by the Copyright Licensing Agency.
Enquiries concerning reproduction outside those terms should be sent to the publishers.

© Springer-Verlag London Limited 1992
Printed in Germany

The publisher makes no representation, express or implied, with regard to the accuracy of
the information contained in this book and cannot accept any legal responsibility or
liability for any errors or omissions that may be made.

69/3830-543210 Printed on acid-free paper

To Adriana, Fiona and Karla for her love, patience, support and understanding during all these years and while I was writing this book.

To the memory of my father, to my mother and all my family, for their love.

To my close friends, in appreciation of their friendship and unconditional support.

Fabian

SERIES EDITORS' FOREWORD

The series *Advances in Industrial Control* aims to report and encourage technology transfer in control engineering. The rapid development of control technology impacts all areas of the control discipline. New theory, new controllers, actuators, sensors, new industrial processes, computing methods, new applications, new philosophies, ..., new challenges. Much of this development work resides in industrial reports, feasibility study papers and the reports of advanced collaborative projects. The series offers an opportunity for researchers to present an extended exposition of such new work in all aspects of industrial control for wider and rapid dissemination.

Over the last 20 years advanced control techniques for industrial applications have been developed which are computer friendly and are both flexible and robust. A common feature of these algorithms is that they require greater computing power and may need to differentiate between background and foreground processing. It is fortunate that the technology of control systems has progressed at the same rate as the theory has developed. Parallel processing methods and transputers can provide the additional computing power necessary to implement sophisticated modern control algorithms both for signal processing and control. In this text, Dr. Garcia-Nocetti and Professor Fleming offer an introduction to this area which should be particularly valuable to engineers in industry and research engineers concerned with the advanced implementation of digital controllers.

The text covers both the software for online control purposes and the control design software needed for controller derivation. The MATLAB environment, which is now wide spread, is employed and a variety of industrial applications are described. The main aim of the text is to alert the reader to the opportunities provided by use of such techniques and to the theoretical challenges the technology provides. It also provides a view of a computing environment which is likely to impact on many future areas of real control applications.

We are delighted to have this Volume in our Series, a volume which is nicely complemented by Dr. Thompson's contribution to the AIC Series *Parallel Processing for Gas Turbine Control*.

M.J. Grimble and M.A. Johnson
Industrial Control Centre
Glasgow, Scotland, UK

PREFACE

Rapid developments in computing technology over the last few decades have had an important impact on controller implementation. The emphasis now is very firmly on the use of digital control to take advantage of its attendant versatility and reliability. Control engineers quickly adopted the use of general-purpose microprocessors in the 1970s enabling more detailed control system designs to be conceived. The growing complexity of digital control systems, in such areas as robotics, flight control and engine control, has created a demand for even faster and more reliable systems. This book examines how parallel processing can satisfy these requirements.

Following a general introduction to parallel processing architectures, multiple instruction multiple data machines (MIMD) machines are identified as suitable systems for tackling digital control problems, which are characterised as a mixture of regular and irregular algorithmic tasks. The Inmos transputer, together with its associated parallel programming language, occam, is introduced as a suitable target system for real-time control.

Confronted with a possible parallel processing implementation, the control engineer encounters a new problem associated with the efficient mapping of real-time control tasks onto individual processors of the system. The authors have experience of work on a variety of industrial applications (flight control, radar tracking, gas turbine engine control, etc.). As a result, a variety of mapping schemes which are described and assessed, help to illustrate potential areas of difficulty. Solutions are proposed and tested on a flight control case study which is used as a theme example throughout the book.

An important implementation tool, EPICAS (Environment for Parallel Implementation of Control Algorithms and Simulation) is presented. This tool, founded on familiar computer aided control system design (CACSD) software, enables the rapid development, analysis and implementation of real-time control laws. Recognising the widespread acceptance of MATLAB and its derivatives for CACSD, it is demonstrated how mapping strategies can be realised in this environment and, further, integrated with a transputer development system for on-line performance evaluation. Examples of use of this tool affords the opportunity to study and analyse important issues associated with parallel implementations of digital control. Mapping strategies, topological considerations and appropriate task size are all investigated in depth.

The authors hope that this book will help introduce the reader to some key topics in this exciting area of development in digital control and will encourage further work in application areas and the generation of new support tools.

D.F. Garcia Nocetti
P.J. Fleming

ACKNOWLEDGEMENTS

The authors wish to acknowledge the help of the following people:

Gordon Ingle (RAE, Bedford) for providing us with the flight control law description and, subsequently, for many interesting discussions, together with Eddie Bailey (RAE, Bedford).

Prof. B.A. White (RMCS) and Dr. D.I. Jones (University of Wales, Bangor) for many helpful suggestions in the early stages of constructing this book.

Many colleagues (University of Wales, Bangor) involved in parallel processing applications -in particular, Dr. H.A. Thompson, Dr. D.I. Lawrie, Dr. P. Entwistle and C.M. Jones.

Prof. J.J. O'Reilly, Head of School of Electronic Engineering Science, for his encouragement at University of Wales, Bangor, and support of research in this area.

CONACYT (Consejo Nacional de Ciencia y Tecnología), Universidad Nacional Autónoma de México (UNAM) and ORS Awards Scheme for the support of Dr. D.F. Garcia Nocetti in his research work at University of Wales, Bangor.

CONTENTS

1 INTRODUCTION 1

    1.1   THE NEED FOR PARALLEL PROCESSING 1
    1.2   OUTLINE OF THE BOOK 2

2 PARALLEL PROCESSING AND THE TRANSPUTER 6

    2.1   INTRODUCTION 6
    2.2   PARALLEL PROCESSING COMPUTER ARCHITECTURES 8
    2.3   WHY USE OCCAM AND TRANSPUTER ? 12
    2.4   THE TRANSPUTER 14
    2.5   PARALLEL PROGRAMMING AND OCCAM LANGUAGE 18
    2.6   REAL-TIME CONTROL TRANSPUTER APPLICATIONS 25
    2.7   SUMMARY 28

3 PARALLEL ALGORITHM STRATEGIES 30

    3.1   INTRODUCTION 30
    3.2   PARTITIONING AND TASK ALLOCATION 30
    3.3   VAP CONTROL LAW 31
    3.4   MAPPING STRATEGIES 33
    3.5   SUMMARY 50

4 EPICAS - AN ENVIRONMENT FOR PARALLEL IMPLEMENTATION OF CONTROL ALGORITHMS AND SIMULATION 51

    4.1   INTRODUCTION 51
    4.2   EPICAS 51
    4.3   MATLAB TOOLS FOR PARALLEL PARTITIONING 53
    4.4   OCCAM TOOLS FOR TASKS ALLOCATION 63
    4.5   USING EPICAS FOR SIMULATION 83
    4.6   SUMMARY 89

5 PERFORMANCE ISSUES: GRANULARITY, TOPOLOGY, MAPPING STRATEGIES 91

    5.1   INTRODUCTION 91
    5.2   MEASURING PARALLEL PROCESSOR PERFORMANCE 91
    5.3   PROCESSOR FARM PERFORMANCE 96
    5.4   PROCESSOR STAR PERFORMANCE 108

5.5	GRANULARITY ISSUES	111
5.6	SUMMARY	130
6	**CONCLUDING REMARKS**	**132**
6.1	GENERAL REVIEW	132
6.2	FUTURE WORK	134
6.3	SUMMARY	138
REFERENCES		139
INDEX		145

CHAPTER 1

INTRODUCTION

1.1 THE NEED FOR PARALLEL PROCESSING

Typically, digital control software consists of real-time control and identification functions together with a number of other activities related to event logging, checking, database management and input/output handling. In general, these operations must be performed within a certain sample time.

Due to growing system demands, many modern controllers need a powerful computing capability to achieve their required performance. For example, new airframe designs and extended aircraft performance envelopes require more complex control laws to be computed at high sampling rates in order to be able to perform them within a fast real-time control loop [1]. The control of electric motors demands short sample intervals because of the small time constants involved and requires control calculations to be carried out in a relatively short interval. Multivariable systems add complexity to the control calculations since several control signals must be calculated simultaneously. Where reliability is essential, for example, in aerospace and nuclear control applications, incorporation of software fault tolerance features adds considerable software overheads. In recent years, therefore, in many instances, digital control requirements have begun to outstrip the performance of conventional general purpose digital computers [2].

This has stimulated control engineers into investigating the potential of novel architectures for low-cost high-speed computation. The availability of a range of parallel processing computer architecture designs, in particular, is creating new opportunities for the implementation of real-time control functions, allowing faster systems to be controlled and giving the control engineer the choice of added complexity in the control algorithm [3], [4]. In particular, the emergence of the INMOS *transputer* and the *Occam* language, for the support of parallel processing has prompted considerable research into real-time control applications [5].

Parallel processing using transputers offers a potentially cost effective solution. The "network" configuration supported by the transputer and its special parallel programming language, Occam, are contemplated as suitable elements for the support of parallel processing for real-time applications. This is based on the suitability of Occam as a language for programming real-time control systems, and the potential of the transputer as a computing component in building high performance controllers. The solution is cost effective since the scale of the hardware solution may be matched to the requirements.

1.2 OUTLINE OF THE BOOK

Chapter 2 reviews various parallel processing architectures commenting on their relevance for real-time control. The INMOS transputer and its special parallel programming language, Occam, are introduced and considered as suitable elements for the support of parallel processing in real-time control applications. The Chapter concludes with a selective review of parallel processing applications for real-time control.

A number of parallel processing mapping strategies are described in **Chapter 3** and their effectiveness is assessed. Suitable techniques are identified for the implementation of real-time control algorithms on concurrent processors. The strategies presented include the Heuristic approach and the Parallel Branches approach, two techniques developed in [6]. New strategies include the Hybrid and Parallel State-Space approaches and a method based on the Factored State Variable Description [7]. All these methods are assessed with respect to execution speed, ease of programming and adaptability.

A number of factors has influenced the achieved speedup and efficiency of the implementation. These factors are task precedence within the algorithm, the need to communicate data between tasks, and the effect of unequal task lengths. The strategies allow the control engineer to perform the partitioning of the control algorithm as a number of tasks which are statically allocated onto the target architecture. This requires a careful partitioning of the algorithm into a group of tasks and an organised allocation of the workload on the parallel processing system.

The effort involved in performing partitioning and task allocation by hand, has led to the development of software tools to automate the implementation process. We have integrated, in a PC environment, the control system design package, MATLAB [8], with a Transputer Development System (TDS), to generate **EPICAS** (an Environment for Parallel Implementation of Control Algorithms and Simulation). This is described in **Chapter 4.**

This environment offers the control engineer a number of software tools for automating the implementation of control algorithms and simulation systems on transputer-based architectures. Within this environment, a *MATLAB-based toolbox* has been created to provide utilities to automate the parallel partitioning of a given system. These partitioning tools generate a parallel representation of the original control system as a number of independent tasks. As practical control systems are often described in block diagram form, the software tools facilitate the entry of block diagram system descriptions. Systems containing nonlinear elements have also been accommodated.

The software performs the partitioning, discretisation and parallel representation of the system as a number of first- or second-order state-space equations. An *Occam-based toolbox* has been developed to automate the mapping of the MATLAB partitioned representation of the system tasks onto a number of transputer-based topologies, using either static or dynamic task allocation strategies. The utilities included in these tools permit the evaluation of the allocation strategies, by displaying on-line task allocation, processor activity and execution time.

The MATLAB and Occam toolboxes developed within EPICAS are described and illustrated using a theme example - the VAP flight control law - which is maintained throughout the book. First introduced in **Chapter 3**, the Versatile Auto-Pilot (VAP) control law (a 4-input, 2-output control algorithm) is one of two control laws for approach and landing developed in theoretical studies at Royal Aerospace Establishment, Bedford, [9], [10]. The control law had been previously been flown on the Civil Avionics Section's BAC 1-11 using a single-processor (M68000) implementation. A 4 transputer implementation has subsequently been flown as part of a Demonstrator project.

An example of using the toolboxes for simulation purposes is also illustrated in **Chapter 4** and serves to demonstrate the wide applicability of the approach to both real-time control and simulation. A transputer-based implementation of the models of the flight dynamics of a BAC 1-11 transport aircraft is generated in order to observe its response to selected test scenarios.

Chapter 5 analyses the performance benefits of the task allocation strategies running on a number of multiple processor structures. A variety of metrics are used for measuring the performance of the strategies. A partitioned version of the digital flight control law has been mapped onto a number of parallel processing transputer-based systems. This particular theme example (of fixed size) has provided insight into the potential performance improvements and bottlenecks of the target systems. However, the results can vary strongly with the algorithm and size of the problem. This consideration leads us to apply the software tools to more complex problems. Models with a larger number of tasks, and higher task granularity have

Introduction *Chapter 1*

been considered. This permits the analysis and evaluation of the performance of the tools in a more general manner.

Chapter 6 closes the book with a review of the key issues and suggestions for further work to make parallel processing more accessible to the practising control engineer and to further strengthen its appeal.

CHAPTER 2

PARALLEL PROCESSING AND THE TRANSPUTER

2.1 INTRODUCTION

A general parallel processing system is composed of several processing elements (PEs) which can operate concurrently, communicating with each other when necessary, see Figure 2.1. Parallel architectures differ both in respect of the processing power of their PEs and the degree of interconnectivity between them.

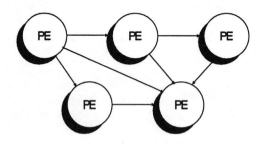

Figure 2.1 Generalised parallel processing system.

This leads to issues such as task granularity, which is a measure of the size of the tasks that can be effectively executed by the PEs of a specific architecture. PEs of fine-grain architectures are characterised by having limited functionality and a wide bandwidth for local data communication. On the other hand, PEs of coarse- or medium-grain architectures are more general-purpose and the interprocessor communication bandwidth is narrower [11].

We have already noted, in Chapter 1, that as real-time applications become more complex, in an increasing number of applications, the general-purpose microprocessor is unable to execute the necessary operations in the required time. The potential speedup offered by a parallel processing system, therefore, suggests a solution to both the speed and complexity problem. Other benefits also arise out of the use of parallel processing. At first sight, a parallel processing system, consisting of a number of PEs (processing elements), is likely to lead to reduced reliability. However, the flexibility of the system, its connectivity and ability to reconfigure, afford an opportunity for alternative system architecture designs to improve reliability.

Further, recognising that concurrent operations are naturally occurring phenomena in real-time systems, the use of parallel programming enables the software designer to more clearly express the variety of sequential and parallel tasks to be undertaken by a digital system. Finally, the scalability of a multiple processor system, properly designed, permits easy expansion to accommodate growing requirements.

Parallel processing architectures are classified in the following Section. A range of architectures is described, and comments are made on the relevance of each architecture for real-time systems. The Inmos transputer has been readily accepted by the control engineering community as a most suitable computing element available for embedded parallel processing systems. Its architecture is described and its associated parallel programming language, occam, is introduced in some detail, paying attention to real-time issues. The Chapter closes with a look at some applications of parallel processing in real-time control.

2.2 PARALLEL PROCESSING COMPUTER ARCHITECTURES

Conventional programming languages such as Pascal and C operate in a sequential manner, where one program statement is executed at a time. This sequential nature has been forced by the sequential architecture of conventional computers (based on von Neumann's concepts), using a single processor, memory and input/output devices linked via a single data bus, see Figure 2.2. However, many problems such as vision, speech, simulation, digital signal processing and digital control have an inherent parallelism and would be better solved in parallel mode.

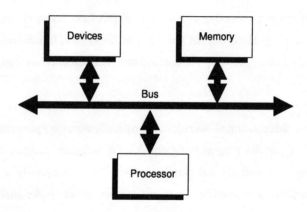

Figure 2.2 Block diagram of a conventional sequential computer system.

The algorithms for computing the solutions of these problems would need to be stated in parallel terms, and have a suitable means (programming language) of expressing parallelism. Parallel architectures are classified according to a number of different criteria.

One of the most widely used classification systems was introduced by Flynn [12], see Figure 2.3, who considered the traditional sequential von Neumann model as a single stream of instructions operating on a single stream of data (SISD).

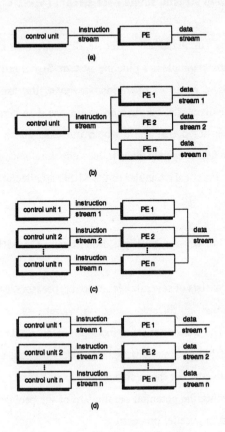

Figure 2.3 Flynn's classification.

According to Flynn's taxonomy, parallel architectures are:

i) **Single Instruction Stream, Multiple Data Stream (SIMD).**

The same instruction is broadcast to all PEs which will execute this instruction simultaneously on different data. Array processors are an example of this architecture and are well suited for implementing regular algorithms involving matrix operations.

ii) **Multiple instruction stream, single data stream (MISD).**

Several processors simultaneously execute different instructions on a single data stream. One possible example is a pipeline system. Signal processing algorithms often exploit the pipelined mode of operation, however, the user must be wary of the latency interval involved, i.e. the time taken for one data item to propagate along the pipe. (It should be noted that some would question that a pipeline is an accurate representation of a MISD system since it does not simultaneously operate on the same "single data" item. Practical examples of the MISD architecture, interpreted in its strict sense, are rare.)

iii) **Multiple instruction stream, multiple data stream (MIMD).**

This architecture consists of several independent processors, each capable of executing different instructions on different data. A variety of topologies exist for the interconnections. The transputer is an example of a MIMD architecture. Compared with SIMD and MISD architectures, MIMD machines can, in principle, deal with a wider range of problems but the designer must play a direct and fundamental role in successfully extracting the potential parallelism of the problem and in evaluating the trade-offs involved in parallel processing.

It should be noted that, while useful, this classification is by no means exhaustive. In general, currently available SIMD architectures are fine-grained, with individual tasks consisting of a single operation, whilst MIMD architectures are usually medium- or coarse-grained.

The issues involved in the parallel implementation of an algorithm are directly dependent on the target architecture, the choice of which should itself be a function of the problem under consideration, as different problems display different levels of parallelism to be exploited. MIMD machines are deserving of special consideration due to their flexibility and ability to operate on unstructured and unpredictable operations and data. While they offer versatility to parallel computing they are also a difficult architecture class with which to work.

Performance benefits strongly depend on the compute/communicate ratio [13]. This ratio expresses how much communication overhead is associated with each computation. Clearly a high compute/communicate ratio is desirable. The concept of task granularity can be also viewed in terms of compute time per task. When this is large, it is a coarse-grain task implementation; when it is small, it is fine-grain. Although large grains may ignore potential parallelism, partitioning a problem into the finest possible granularity does not necessarily lead to the fastest solution, as maximum parallelism also has the maximum overhead, particularly due to increased communication requirements. Therefore, when partitioning the application across PEs, the designers must choose an algorithm granularity that balances useful parallel computation against communication and other overheads.

MIMD machines need not be connected in any specific way; example topologies are given in Figure 2.4. While bus interconnection (a) offers the simplest solution, it has the highest possible contention (competition for communication channels). Conversely, the crossbar configuration (b) offers the least contention, but has the highest complexity.

In this book we concentrate on medium-grain MIMD architectures, because these offer more flexibility for handling irregularly structured general purpose software. Numerous processors can be linked together in a loosely-bound network in which they all have their own independent memory, or they can be combined as tightly-bound multiprocessor in which each processor can access any memory.

A range of medium-grain architectures based on a range of topologies have been developed. These include the COSMIC cube [14], the IBM RP3 [15], transputer-based systems, etc. Our main interest is MIMD system implemented with transputers programmed in occam language.

Parallel processing and the transputer   *Chapter 2*

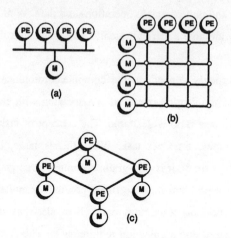

Figure 2.4 MIMD processor-memory configurations:
(a) Shared bus; (b) Crossbar; (c) Point-to-point.
P,PE = Processor Element; M = Memory.

2.3 WHY USE OCCAM AND TRANSPUTER ?

Attempts have been made to extend the simple von-Neumann design to multi-processors for parallel operation using a single-bus for MIMD machines, see Figure 2.5. However, a number of problems arise when using a single-bus multiprocessor. Adding more PEs to the data bus, after an initial improvement, causes a degradation in the overall performance, due to the increased competition for use of the shared data bus (bus-contention), resulting in extended "idle" times for each extra processor as they wait to gain access to the bus since in the von Neumann model only one processor may have access to the bus at any time - a condition known as the von Neumann bottleneck [16].

Software creates another problem. A number of approaches to parallelism in software have been considered in the context of a multi-processor environment. The programming languages used for these systems were, by and large, extensions of existing sequential-mode languages (e.g. Parallel C, Parallel Fortran) for providing parallel programming features. Any program

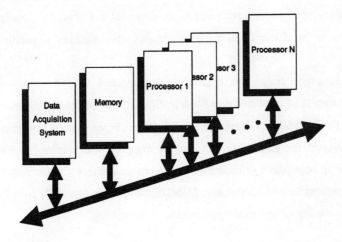

Figure 2.5 Single-bus multiprocessor configuration.

written for a multi-processor system must be partitioned across all the processors into self-contained sections which then will then run concurrently, communicating with each other when necessary. Conventional programming languages, however, were not initially designed to cope with these issues. A better solution is clearly a high-level programming language in which concurrency and efficient intercommunication are directly supported. Much current research however is being devoted to the design of new parallel computer architectures and corresponding formalisms for programming languages. The *occam* language and the *transputer* architecture represent one of the success stories.

Occam is an abstract language that has the dual role of being an implementation language and a design formalism. It has the ability to control and synchronise processes via channels on one or more processors. It was developed specifically for parallel programming, based on a model of concurrency and communication that match to the transputer. Occam programs which have been designed for a multi-processor environment can be written and tested on a single processor and then transferred to the multi-processor environment. Modula-2, Linda and Ada are other examples of concurrent languages. However their model of concurrency has not been implemented as efficiently as the occam model when executed in parallel [17].

The occam language design has been closely associated with the development of the transputer. Occam is an ideal language for scientific and engineering applications, for control applications, for embedded systems, and for simulation. When used on transputers it provides operational support for scheduling and process mapping with minimal overhead. Parallel processing using transputers programmed in occam also offers a potentially cost effective solution, when compared with alternative MIMD architectures, especially suited to the support of parallel processing of real-time applications.

A transputer is a programmable VLSI device containing a processor, local memory and communication links for inter-connection with other transputers. One transputer used alone is a high-performance microprocessor. When it is required to exploit concurrency, transputers can be collectively configured in a network to build a higher-performance and non-contention concurrent system, see Figure 2.6. Transputer hardware supports the occam model of concurrency, which provides a framework for designing concurrent transputer-based systems programmed in occam. Occam's model of concurrency has been influenced by the need to provide the same programming techniques on a single transputer and a network of transputers.

2.4 THE TRANSPUTER

The transputer (contraction of the words: *transistor* and *computer*) is a new generation VLSI architecture which explicitly supports concurrency and synchronisation [18]. The transputer is a family of single-chip computers which incorporates features to support the occam model of parallelism.

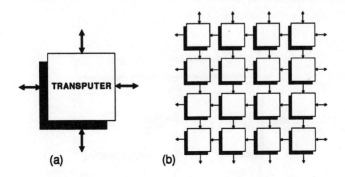

Figure 2.6 (a) A transputer. (b) A transputer network.

2.4.1 Transputer Architecture

The T800 version of the transputer is shown in Figure 2.7. It consists of a 32-bit 10 MIPS RISC processor, 4Kbytes of local RAM, an external memory interface and four link interface units. The point-to-point configuration is supported by the Inmos transputer through these four link interface units per PE to facilitate interprocessor communication [19],[20]. These links permit the transputer to be configured as arrays of processors, thus avoiding the potential communication bottleneck of a single-bus system, the complexity of the crossbar configuration and the delays associated with the ring topology. Bit-serial communication rates of 20Mbit/s are achieved via the four bidirectional interprocessor links. Serial/parallel data conversion is performed by the on-chip link interface hardware. The simple machine structure of the Transputer allows the complex instructions necessary for occam's process handling and message passing to be implemented in microcode. It uses different procedures for external channel communication (via the links) and internal channel communication (on-chip interprocess communication). Provided that frequently accessed data can be held in local RAM, a high level of performance is maintained by the stack-oriented six-register architecture. We will consider transputer implementations of real-time systems later and also study the performance of transputer-based architectures.

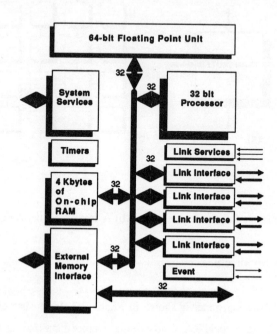

Figure 2.7 T800 transputer architecture.

2.4.2 Support for Concurrency

There are two main approaches to memory access within a MIMD system: *shared memory* and *local memory*. The shared memory approach raises problems associated with maintaining data integrity when data items are accessed by more than one processor. A number of architectural solutions adopt a hybrid approach which uses a mixture of shared memory and local memory. The transputer architecture is based on the local memory model. Thus in a multi-transputer configuration, a transputer has sole use of its own on-chip and off-chip memory and does not have to compete with other transputers for memory access, for instructions and data on a shared data bus. Consequently networks of transputers should scale linearly in performance according to the number of transputers in the network. There is no von Neumann bottleneck to degrade performance. Groups of parallel processes comprising an occam program may be distributed over such a network of transputers. Each transputer will execute its own process -any communication between processes on different transputers being handled by the link -and so the parallel execution of the program will be effected. The normal method of program development using transputers would be to design, implement and test the occam program on a single transputer system, and then when satisfied, to distribute the component processes to the transputers in the network. This requires configuring or mapping processes to the transputers in the network - declaring which processes will execute on which transputer. The close association of the transputer and the occam language has produced efficient parallel processing systems.

2.4.3 Transputer-based Topologies

Different network topologies may be created with a system of transputers by connecting up the transputer links in different ways. These topologies range from pipelines, through rings, to hypercube, as is shown in Figure 2.8.

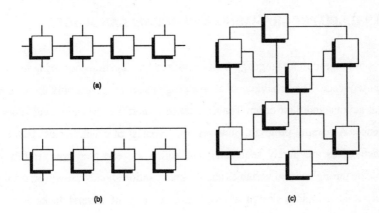

Figure 2.8 Examples of Transputer Topologies: (a) pipeline; (b) ring; (c) hypercube.

Much of the interest in transputer networks results from the fact that they are readily reconfigured. A new generation of transputer systems, T9000 series, is under development [21]. A major goal for this new type of transputer is to provide a significant increase in the performance of the transputer, whilst maintaining instruction set compatibility with previous devices. The processor uses a pipelined architecture, which will be able to execute up to 8 instructions on each clock cycle, operating at a clock speed of 50 MHz. It will provide performance exceeding 150 MIPS and 20 MFLOPS. As the flow of multiple instructions through the pipeline is controlled by hardware, it is not necessary for the existing compiler to be modified, or for source code to be recompiled, to obtain the full claimed performance. There will be 16 Kbytes of on-chip cache memory, and, importantly, a radically new communications system. The T9000's links which will operate at 100 Mbits/second providing a bidirectional communication rate of 20 Mbytes/second, the total bidirectional bandwidth being 80 Mbytes/sec. A link converter chip will allow T9000 links to be connected to T series links and there will be a universal link adaptor.

While the transputer was designed specifically to support occam, it is possible to run such languages as Parallel-C, Pascal and FORTRAN. In the next Section the occam programming language associated with the transputer is described.

2.5 PARALLEL PROGRAMMING AND OCCAM LANGUAGE

A sequential programming language is characterised by its actions occurring in a strict, single execution sequence. The behaviour of such a program thus depends only on the effects of the individual actions and their order. The time taken to perform the individual actions is not of consequence. A parallel program, however, may consist of a number of tasks or processes which themselves are purely sequential, but which are executed concurrently and which communicate through shared variables and synchronisation signals. However, even on a single processor, it is useful to design a program consisting of several tasks intended to run concurrently, thus representing more accurately the concurrent nature of many real-world problems. Here, the illusion of concurrency is created through time-sharing the various tasks.

Two important issues which must be addressed by a parallel programming language are those of synchronisation and the handling of shared variables. When several tasks are executing concurrently, they do so asynchronously, i.e. each task proceeds at its own speed. If the activities of these tasks are dependent on one another then the programming language must provide a means of synchronisation to co-ordinate their activities. Further, when two or more tasks are accessing the same variable, control must be exercised to prevent data corruption arising from conflicting operations on that variable.

A number of mechanisms has been devised to handle these two issues - synchronisation and shared variables - such as semaphores, guards, data monitors, rendezvous, etc. However these mechanisms requires careful structuring of the software by the programmer. Through its message-oriented approach, based on "processes" and "channels", occam gives a higher level of support to the programmer and reduces the overhead associated with concurrent programming design.

2.5.1 Occam Model

The basic unit of occam programming is a "process", which performs a set of operations and then terminates. This concept is similar to that of sequential programming languages except that, in this case, more than one process can be executing at any time. Communication between processes is via point-to-point links known as "channels".

Occam does not permit the use of shared variables. Instead, shared data is communicated to processes across the communication channels. Synchronisation is handled in a natural way through the implementation of input and output commands.

2.5.2 Occam Primitives

Processes in occam are built from three "primitive" processes:-

assignment   v:=e

where the variable "v" is set to the value of the expression "e",

input   c?v

where a value is sought from a channel "c" and is stored in the variable "v", and

output   c!e

where the value of the expression "e" is output to the channel "c".

When an output command is encountered in a process, the process is halted until another process has executed the corresponding input command. Then the desired data communication takes place over the channel. Similarly, an input command cannot be executed until the corresponding output command is reached in another process. In this way, communication cannot proceed until both processes are ready to perform the I/O transfer and synchronisation is thus secured.

2.5.3 Occam Constructs

Several primitive processes can be combined into a larger process to form a construct, which is itself a process and may be used as a component of another construct. These constructs are:

- **SEQ**   -sequential construct,
- **PAR**   -parallel construct, and
- **ALT**   -alternation construct.

Typical constructors such as **IF** and **WHILE** are also supported. Since occam is a concurrent programming language, sequential, parallel or alternation execution must be specified.

Parallel processing and the transputer *Chapter 2*

The SEQ construct signifies that the statements inside the process are to be executed sequentially. In the following code example, a measurement signal is input from an ADC, scaled and the result output to a DAC:

```
SEQ
  ADC? meas.signal
  scaled.signal := meas.signal * scale.factor
  DAC! scaled.signal
```

Note that indentation is used to indicate program structure. Here it is used to denote that the three operations following SEQ are components of that process. When two or more operations are to run in sequence, then the sequential mode, SEQ, must be explicitly requested. Figure 2.9 illustrates the relationship between processes and channels.

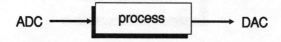

Figure 2.9 SEQ construct example.

A construct, such as SEQ and PAR, is itself a process and may be used as a component of another construct, thus providing the facility to nest processes and hierarchically decompose complex problems. The folding editor contained in the Occam Programming System (OPS) is a useful tool which supports this feature in an elegant way. Using the folding editor we could represent the sequential process sample above thus:-

    ... process

where ... prefixes the fold "title".

Employing this facility, we can illustrate an example of the use of the PAR construct. This construct signifies that the processes following it are to be executed in parallel:

    PAR
        ... process 1
        ... process 2

Thus we see from Figure 2.10 that we can arrange for two ADC/input - scale - DAC/output exercises to run in parallel. In general, parallel processes run asynchronously and synchronisation is only necessary when processes need to communicate over a channel.

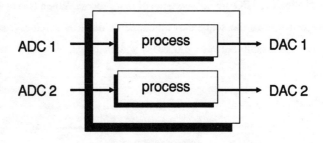

Figure 2.10 PAR construct example.

The ALT construct is an unusual one which often causes initial difficulty. ALT is used where a subset of the input channels may be used to initiate computations within a process, e.g.

    ALT
    input.channel1 ? x
        ... process 1
    input.channel2 ? x
        ... process 2
    input.channel3 ? x
        ... process 3

Parallel processing and the transputer   *Chapter 2*

ALT operates a "first-wins" procedure in which only the process associated with the first input to become ready is executed. In the above example, if input.channel2 was the first to produce an input, then only process 2 would be executed. Each component of an ALT construct, then, begins with a test condition. For example, "input.channel1 ? x" tests whether input.channel1 is ready. If more than one guard is TRUE, then any one but only one of the processes governed by a TRUE guard may be executed. When this situation arises, selection of the guard to be executed is arbitrary.

2.5.4 Real-Time Programming Issues

Prioritization facilities are essential in a real-time programming environment. Occam provides the means to prioritise both PAR and ALT constructs. The PRI PAR construct is especially useful in applications where it is important to service a specific request promptly:

```
PRI PAR
    SEQ
        alarm ? any
        ... service alarm
    SEQ
        ... another process
```

Here, should any contention arise, the process which services the alarm signal has priority over any other process. Possible contention exists when the number of channels ready to receive input exceeds the number of available processors. In the following example, a PRI ALT construct is used to terminate a WHILE loop:

```
SEQ
    running:= TRUE
    WHILE running
        PRI ALT
```

```
            exit ? any
                running:= FALSE
            ch1 ? x
                ... process 1
            ch2 ? x
                ... process 2
```

If a standard ALT construct had been used here, we would have no guarantee that channel "exit" would ever be read and therefore that the loop would be completed. Further details of occam programming may be found in [22].

Current transputer hardware provides very fast pre-emptive scheduling for two static priority levels, with "round-robin" management within each level. However, only two levels of priority can often prove to be insufficient in real-time control applications. Bakkers and van Amerongen [23], for example, suggest that real-time control software might reasonably contain five priority levels ranging from high priority, pre-emptive to a low level implemented by the transputer's low-priority, round-robin scheduler. These are not accommodated within the basic transputer-occam combination.

A number of real-time kernels have been proposed, such as Trans-RTXc [24]. In an attempt to retain efficiency, Welch [25] has proposed two schemes, implemented in occam with an acceptably low level of overhead, to realise multiple, fixed priority or multiple, dynamic priority scheduling on transputers.

2.5.5 Process Mapping and Task Allocation

So far we have referred to "processes" in connection with occam program construction and not to "processors". The latter term is the physical division of a task related to the target hardware while the former term is the logical division of the task. Occam provides a way to allocate these logical processes to actual hardware processors via **PLACED PAR** statements.

The number of processors need not necessarily match the number of designated processes since it is possible to allocate several processes to one processor. It is good practice to decompose the problem to a greater, rather than lesser extent, creating more logical processes than physical processors. It is easier to group processes rather than sub-divide them once a design is established.

There are two main approaches to allocating tasks to processors: *statically* and *dynamically*. In static allocation, the association of a group of tasks with a processor is resolved before running time and remains fixed throughout the execution, whereas in dynamic allocation, tasks are allocated to processors at running time according to certain criteria, such as processor availability, intertask dependencies and task priorities. Whatever method is used, a clear appreciation is required of the overheads and parallelism/communication trade-offs already mentioned. Dynamic allocation offers the greater potential for optimum processor utilization, but it also incurs a performance penalty associated with scheduling software overheads and increased communication requirements which may prove unacceptable in some real-time applications. The effect of such task allocation strategies, with respect to control and simulation algorithms, is studied in more detail in Chapter 5.

2.6 REAL-TIME CONTROL TRANSPUTER APPLICATIONS

We conclude this Chapter with a selected review of some applications of transputers in real-time control.

2.6.1 Robotics

Robot control functions range through a number of levels: from decision making, path planning, and coordination at the top level to joint angle control at the bottom level. The computational requirements of high speed, high bandwidth, adaptive systems makes this area ripe for exploitation by parallel processing. Indeed, the variety of computational tasks suggests the use of a mix of architectures of differing granularity. In particular, parallel processing

Parallel processing and the transputer   *Chapter 2*

strategies have been developed for path planning, coordination and joint control centred on the use of transputers.

Mirab and Gawthrop [26] review work on the application of parallel processing for calculation of the dynamic equations of robotic manipulators. Jones and co-workers [27], [28] report on the use of transputers in this respect and suggest a granularity mismatch, this particular problem being more suited to a fine-grain architecture. Their work is further aggravated by the use of a processor farm topology for which they propose hardware improvements. (The processor farm topology is addressed in detail in Chapters 4 and 5 of this book.)

Daniel and Sharkey [29] draw attention specifically to the heavy computational demands of force control and to the need for a controller which is capable of switching easily between different layers in the control hierarchy. It is argued that latency is a key factor in the determination of a suitable transputer-based system architecture and they advocate their "Virtual Bus" solution.

2.6.2 Motor Control

Until recently, the DC motor has predominated in high performance applications where a fast change of torque is required. The computing load demanded by AC machines has prohibited their application in this domain. Jones et al. [30] and Asher and Sumner [31] have investigated whether transputer-based systems can provide sufficient real-time control computing power for the AC induction motor to be an attractive alternative to DC machines. Different control schemes based on the application of functional decomposition are described in [32]. The relatively slow interprocessor communication of the transputer, however, limits performance although its ease of implementation affords the designer valuable insight into the potential of parallelism. It is inferred that a parallel processing scheme is likely to yield a viable solution. Perhaps the new generation of T9000 transputers with planned order of magnitude faster link communications will fulfil this promise.

2.6.3 Kalman Filtering

Growth in complexity of tracking problems has led to parallel implementations of signal processing algorithms, such as Kalman filters. The regular nature of these problems make them contenders for fine-grain implementations as well as medium-grain (e.g. transputer-based) solutions [33]. Alternative architectures such as systolic arrays and PACE are discussed in this context by Lawrie et al. [34].

Parallel processing has been used to effect in a number of Kalman filter applications. Bahramparvar and Gray [35] have devised a transputer-based instrument for testing the surface quality and structural integrity of metal tubes and wires, using this approach. The number of processors (or Kalman filter solutions) required, is determined *a priori* by the number of fault hypotheses to be tested.

Atherton et al. [36] employ probabilistic techniques operating on the output of a set of Kalman filter algorithms running on separate processors to track multiple targets arising from radar measurements. When targets are in the vicinity of one another (for example, when they cross) multiple tracking algorithms must be activated to monitor track development. An interesting problem, here, concerns the optimum number of processors required, since the maximum computational load is dependent on the worst-case number of simultaneous target crossings.

2.6.4 Flight Control

In an early Demonstrator Project, in collaboration with Royal Aerospace Establishment, Bedford, Fleming et al. [37] explored software and hardware mapping strategies for the implementation of an existing flight control law - the Versatile AutoPilot (VAP). This study inspired the automated control law mapping environment (EPICAS) described in this book. The hardware solution involved four transputers in a "star" master-slave configuration and employed static task allocation. Simple fault scenarios were investigated and occam software solutions proposed. The Project was concluded with successful in-flight testing of the parallel processing flight controller interfaced with VME bus-based hardware onboard the BAe 1-11 test aircraft at RAE, Bedford.

2.6.5 Fault Tolerant Systems

Clearly, a parallel processing system has great potential for fault-tolerance. However, it must be recognised that such a system will tend to be less reliable than a uniprocessor system since it generally has more elements. It only becomes more reliable if it can detect a fault and take corrective action, possibly by circumventing the defective sub-system and reconfiguring its software. Thompson and Fleming [38] use existing fault tolerant techniques to evolve an operationally fault tolerant transputer-based architecture suitable for gas turbine engine control. A system topology, constrained by the dual-lane configuration of gas turbine engine controllers, is devised in a way in which the majority of faults are detected, located and masked by means of a three-way vote, consistent with the conventional triplex approach. They draw attention to limitations of the present transputer generation in devising their scheme.

2.6.6 Other Work

The above review has, of necessity, been brief and selective. It is not intended to be comprehensive; it simply serves to represent a variety of application areas for parallel processing in digital control. Amongst other sources, the reader may wish to access [39] and [40] for more information on applications of parallel processing in control.

2.7 SUMMARY

The nature of advances in VLSI technology has resulted in increased computing power generally being made more available through parallel processing architectures of different types rather than through an increased clock rate in uniprocessor systems. Despite the development of faster processors, the real attraction of parallel processing to system designers is its scalability to meet increasing demands. There is a plethora of engineering application areas. To date, much work in the area has been concerned with the efficient extraction and exploitation of concurrency. There is also a belief that the cost of parallel processing elements may become insignificant, thus encouraging their profligate use in addressing problems of growing complexity. This has stimulated control engineers into investigating the potential of novel architectures for low-cost high-speed computation. At present the availability of a range of parallel processing computer architecture designs has offered new opportunities for the implementation of real-time control functions, allowing faster systems to be controlled, giving the control engineer the choice of added complexity in the control algorithm, and the prospect of handling fault tolerance. In particular, the emergence of the INMOS transputer, as a flexible element for the support of parallel processing for real-time applications, has prompted considerable research into this area of control.

CHAPTER 3

PARALLEL ALGORITHM STRATEGIES

3.1 INTRODUCTION

This Chapter describes a study, undertaken in collaboration with the Royal Aerospace Establishment, Bedford, to map an experimental automatic flight control law onto a parallel processing transputer-based system, programmed in occam. A number of strategies are devised for transforming the existing control law into a form suitable for the exploitation of parallel processing. Five partitioning strategies are presented: Heuristic, Parallel Branches, Hybrid, Parallel State-Space, and Factored State Variable Description. These strategies are assessed with respect to execution speedup and efficiency.

3.2 PARTITIONING AND TASK ALLOCATION

There are three fundamental problems to be solved when considering the implementation of a control algorithm on parallel processing systems:

- Identifying parallelism in the algorithm.
- Partitioning the algorithm into tasks.
- Allocating the tasks onto processors.

In general, the parallelism extracted from the algorithm depends on the approach taken by the control engineer to identify or express parallelism. Partitioning is necessary to ensure that the resulting parallel algorithm is suitable for the target multiprocessor system. Care in task allocation is necessary to achieve good processor utilisation and to optimise inter-processor communication in the target system. The partitioning and task allocation should be designed so as to reduce the parallel execution time on the target multiprocessor system.

There is a number of parameters which have a significant influence on the performance of multiple-processor systems. These include the amount of parallelism inherent in the problem, the method for decomposing a problem into smaller modules or tasks, the method applied to allocate tasks to processors, the grain size of the tasks, sequential execution times, communication and scheduling overheads, the number of processors, the speed of processors and memories, and the topology [41], [42], [43], [44], [45], [46]. Analysing the performance of multiple-processor systems is a complex task, since many factors jointly determine system performance and the modification of some factors affects many others. For example, when the grain size of computations executed on processors is reduced, in order to better exploit the parallelism available in the application, processors complete the computation more rapidly. As a result these communicate over the interconnection network more frequently, thus increasing the communication overhead and in turn slowing down processors. Since the interaction among these various factors is complex and involves many tradeoffs, it is necessary to tune system parameters to achieve peak performance at minimum cost.

3.3 VAP CONTROL LAW

The Versatile Auto-Pilot (VAP) control law is described briefly in this Section. This law is one (the most complex) of two control laws for approach and landing developed in theoretical studies at Royal Aerospace Establishment, Bedford, [9], [10]. The control law had been previously been flown on the Civil Avionics Section's BAC 1-11, using a single-processor (M68000) implementation. The VAP is a 4-input, 2-output control algorithm and it is presented in block diagram form in Figure 3.1.

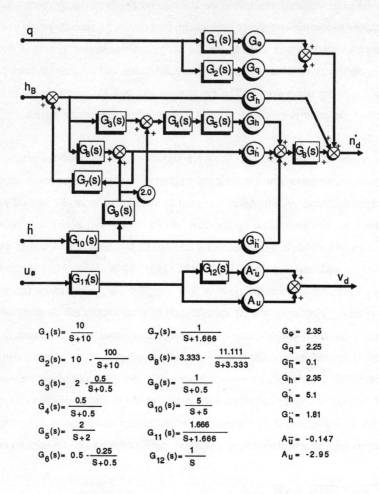

Figure 3.1 VAP control law block diagram.

The inputs are pitch rate (q), barometric height error (h_B), vertical acceleration (d^2h/dt^2) and airspeed error (u_a); the outputs are elevator rate demand ($d\eta/dt$) and throttle demand (v_d). The diagram notation follows the convention that the gain between the elevator position and an aircraft state error input is denoted by **G** with the error variable as a subscript. Similarly, **A** is used to denote the gain associated with throttle position. The throttle position control law consists of a smoothing lag on the airspeed error input together with the **A** gains. As the elevator servo-system is a rate demand servo, the elevator control law demand is of derivative type. The first component of the elevator control law comprises **G** gains together with a 0.1s lag. In the second component of the law, the height error inputs from several sources are mixed in complementary filters with normal acceleration information. These signals are further transformed by lag terms and **G** gains. It should be noted that the VAP control law is not particularly demanding for implementation on a transputer array. However, it is of sufficient complexity, possessing significant cross-coupling terms, to be a non-trivial exercise for parallelisation.

3.4 MAPPING STRATEGIES

This Section describes strategies used to generate the concurrent realisation of the VAP control law on a transputer-based network by means of partitioning the algorithm into a number of tasks and realising a static allocation of the resulting tasks on a number of processors. Five methods have been used: Heuristic, Parallel Branches, Hybrid, Parallel State-Space and Factored State Variable Description (FSVD).

In static allocation, the association of a group of tasks with a processor is fixed. A given group of tasks will always be computed by the same processor. Thus, when partitioning the control law tasks across the available processors, the purpose is to minimise the execution time of the tasks set on the processors. There are a number of interacting factors to be considered in doing so [47]. The main factors are:

- sequentialism within the algorithm or task precedence relationships,
- the need to communicate data between tasks, and
- the effect of unequal task lengths.

A logical strategy is to inspect the controller algorithm for sections where independent tasks exist and to restrict the use of multiple processors to these sections. Common factors within the algorithm are computed but do need to be communicated to the other processors, which introduces a communication overhead. Within the parallel modules, the tasks are allocated so as to balance the computational load across the processors. Clearly there is a loss of efficiency, since some processors will have idle periods. Consequently, the reduction of execution time will fall short of the ideal reciprocal relationship with a number of processors. This is a characteristic of the **Heuristic** strategy.

Another attractive technique, the **Parallel Branches** strategy, has been developed to maximise the parallelism by modifying the algorithm, using block diagram transformation and superposition, into a set of independent tasks. Here, a group of tasks allocated to a given processor must calculate all the processes needed to obtain the result. As there are common factors between the tasks on separate processors, a duplication of effort results. Thus sequentialism is removed at the expense of wasted effort, but the strategy **does** have the benefit of reducing the communication of data between processors to a minimum. The tasks are allocated so as to balance the computational load across the processors, attempting to group tasks with common factors on the same processor. This strategy ensures a high processor utilisation but not necessarily a lower execution time because of the duplicated effort, as will be seen in Section 3.4.2.

In general there is a trade-off between degree of parallelism and speed of communication. These two methods have been reported in [48] and used for comparison purposes here with new techniques. An alternative new approach associated with the **Heuristic** and **Parallel Branches** techniques is a **Hybrid** approach to the concurrent implementation of control algorithms. Using this strategy, we first apply the **Heuristic** technique initially to decompose the problem. Then, the **Parallel Branches** approach is employed to split the critical sequential paths of the control law to execute them concurrently. This technique attempts to exploit the attractive features of both of the original methods.

The **Parallel State-Space** strategy, like the **Parallel Branches** method, seeks to maximise the parallelism by modifying the algorithm into a set of independent tasks. However, this strategy transforms the control law into a number of independent tasks, using a state-space model of the system. This method also generates the relationship between each input and output variable and expands the resulting transfer functions into lower order models, discretising the models and representing them as a number of state-space equations to be executed concurrently.

Information on the precedence of a given control algorithm is useful for a designer considering a parallel implementation. Starting from a **Factored State Variable Description (FSVD)** [7], or its equivalent signal flow graph (SFG), a possible approach is to assign the computation of nodes to different processors in such a way that node computations are executed in parallel whenever possible. This an attribute of the strategy based on the FSVD.

3.4.1 Heuristic Approach

For the **Heuristic** strategy, the discrete-time equivalent of each transfer function block in the original VAP control law was calculated, see Figure 3.2.

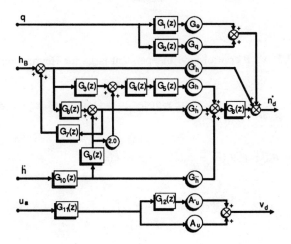

Figure 3.2 VAP control law: Discrete-time model for Heuristic approach.

The dependencies of the blocks along the various paths in Figure 3.2 were inspected. Even though this structure consists of a collection of sequential blocks, parallelism in the model was extracted by inspection of the paths' dependencies and non-dependencies. A number of independent tasks, with no common factors was derived, see Figure 3.3. No further partitioning was viable due to the constraints enforced by sequentialism within the paths.

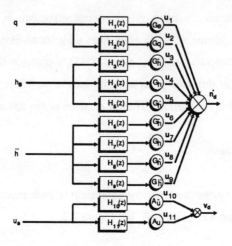

Figure 3.3 VAP control law: Independent tasks for Heuristic approach.

Thus, the outputs of the control law were reduced to the following sequence of computations:

$$\eta_d(k) = u_1(k) + u_2(k) + u_3(k) + \ldots + u_9(k) \quad (3.1)$$

$$\upsilon_d(k) = u_{10}(k) + u_{11}(k) \quad (3.2)$$

where η_d and υ_d are the elevator rate and throttle demands and $u_1, u_2, \ldots, u_{11}$ are the resulting concurrent discrete functions. These discrete functions corresponding to the Heuristic design approach of the VAP control law were mapped on a transputer array, varying the number of

Parallel algorithm strategies   *Chapter 3*

transputers in the array, from 1 up to 6 processors in the test case presented. The software design required careful organization of the discrete-time models into the processes for balancing their intercommunication and the amount of processing performed on each transputer.

3.4.2 Parallel Branches Method

In the **Parallel Branches** method, the control law was modified, using block diagram transformations to generate a parallel network of Laplace transfer functions, as is illustrated in Figure 3.4.

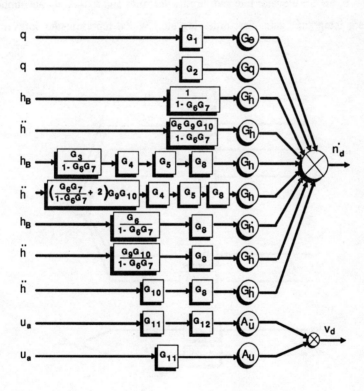

Figure 3.4 VAP control law: parallel paths network.

Then, each of these parallel paths, in the new schematic block description, was expanded into partial fractions and converted to the discrete-time domain, to reduce the control law to a sum of discrete functions. The resulting network of independent blocks for the VAP control law example is shown in Figure 3.5. These blocks are then combined to form the control signals:

$$\eta_d(k) = u_1(k) + u_2(k) + u_3(k) + \ldots + u_{35}(k) \qquad (3.3)$$

$$v_d(k) = u_{36}(k) + u_{37}(k) \qquad (3.4)$$

where η_d and v_d are the elevator rate and throttle demands and $u_1, u_2, \ldots, u_{37}$ are simple discrete functions (e.g integrator, gain, first order lag etc.) which represent the software building blocks.

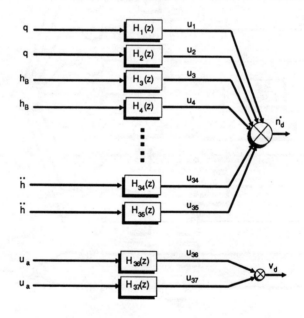

Figure 3.5 VAP control law: parallel branches description.

Parallel algorithm strategies   *Chapter 3*

These building blocks are programmed as individual occam processes and used in a number of occam programs developed for a variety of transputer arrays. In contrast with the Heuristic case, the software design required a simpler organisation of the resulting low-order models, offering a better computational load balance in the final transputer implementation.

3.4.3 Hybrid Method

The **Hybrid** method combines the **Heuristic** and **Parallel Branches** techniques in the concurrent implementation of control algorithms. According to this strategy, we first apply the **Heuristic** technique to initially decompose the problem. Then, the **Parallel Branches** approach is employed to split the critical sequential paths of the control law, to execute them concurrently. The new **Hybrid** representation of the VAP control law is shown in Figure 3.6.

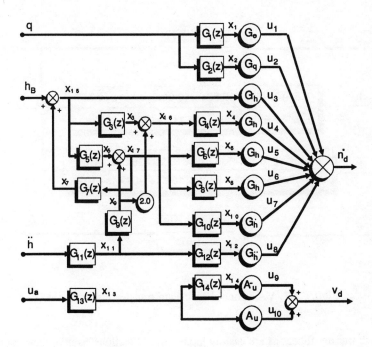

Figure 3.6 VAP control law: Hybrid approach.

Parallel algorithm strategies   *Chapter 3*

In the VAP control law shown in Figure 3.2, there is a number of paths critically long for real-time implementation. An example of these is the path including the sequential blocks $G_4(z)$, $G_5(z)$ and $G_8(z)$. The **Parallel Branches** approach has been used to decompose this particular path into a set of parallel paths to be executed concurrently. The resulting diagram has now a number of new parallel blocks substituting some of the critical paths.

3.4.4 Parallel State-Space Method

The Parallel State-Space method transforms the original control law into a number of independent tasks, by means of obtaining the relationship between each input and output variable, and reducing the control algorithm to a set of independent path transfer functions, as illustrated in Figure 3.7.

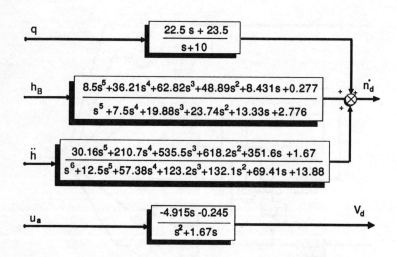

Figure 3.7 VAP control law: Path transfer functions set.

As these path transfer functions are usually higher order functions, each of these path transfer functions was decomposed into a parallel connection structure of lower-order partial fractions

Parallel algorithm strategies Chapter 3

(first and second order). The resulting functions were discretised and finally represented as a parallel set of discrete-time state-space equations, as displayed in Figure 3.8.

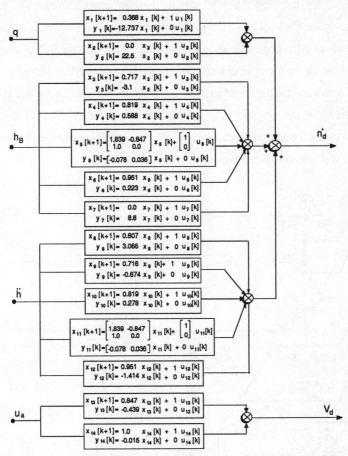

Figure 3.8 Parallel State-Space representation of VAP control law.

This method was developed following experience of using the Parallel Branches method. There are strong similarities between the two approaches. The important difference is that by using the "state-space" approach, tools exist within MATLAB to eliminate redundancies (which persists in the Parallel Branches implementation). This in accomplished by using minimal realisation techniques.

Parallel algorithm strategies   *Chapter 3*

3.4.5 Factored State-Space Description Method

A final method, for mapping the VAP control law on transputer systems, based on the Factored State-Space Variable Description (FSVD) has also been evaluated. A full description of this strategy and its implementation is available in [49]. The FSVD mainly has the ability to partition an input/output description into modules of varying degrees of complexity, and the facility to vary the degree of parallelism of the realization. Using this approach each individual block, in the original diagram description of the VAP, was discretised and represented in direct form. Then, connecting all the submodels together, a SFG representation of the system was generated, see Figure 3.9.

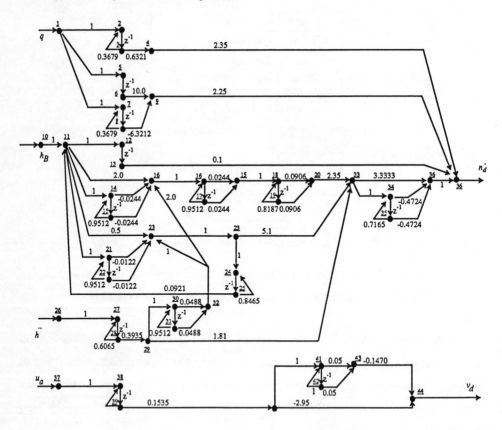

Figure 3.9 VAP control law Primitive Signal Flow Graph.

Parallel algorithm strategies Chapter 3

The construction of a FSVD from a given SFG describing a digital realisation is direct. It involves identifying the order of computation of the intermediate node variables in the SFG, starting with the input and present state nodes and finishing with the updated state and output nodes. Different sets of intermediate nodes are related by products of matrices, and the system can be described in the form:

$$\begin{bmatrix} x_{k+1} \\ y_k \end{bmatrix} = Q_n \times Q_{n-1} \times \ldots \times Q_1 \begin{bmatrix} x_k \\ u_k \end{bmatrix} \quad (3.5)$$

This FSVD preserves all the node variable computations and the order in which they are executed, the number of factors (n) being the length of the longest delay-free path in the SFG. The resulting node computation precedence for the VAP control law is given in Table 3.1.

Precedence Level	Node Computations
I	$N_2 = 0.3679 \times N_3 + N_1$ $N_4 = 0.6321 \times N_3$ $N_7 = 0.3679 \times N_8 + N_1$ $N_9 = 10 \times N_6 - 6.3212 \times N_8$ $N_{11} = 0.0921 \times N_{25} + N_{10}$ $N_{27} = 0.6065 \times N_{28} + N_{26}$ $N_{29} = 0.3935 \times N_{28}$ $N_{38} = 0.8465 \times N_{39} + N_{37}$ $N_{40} = 0.1535 \times N_{39}$
II	$N_{14} = N_{11} + 0.9512 \times N_{15}$ $N_{21} = N_{11} + 0.9512 \times N_{22}$ $N_{30} = N_{29} + 0.9512 \times N_{31}$ $N_{41} = N_{40} + N_{42}$ $N_{12} = N_{11}$
III	$N_{32} = 0.0488 \times N_{30} + 0.0488 \times N_{31}$ $N_{43} = 0.05 \times N_{41} + 0.05 \times N_{42}$
IV	$N_{16} = 2 \times N_{32} - 0.0244 \times N_{14} + 2 \times N_{11} - 0.0244 \times N_{15} + 0.9512 \times N_{17}$ $N_{23} = N_{32} - 0.0122 \times N_{21} + 0.5 \times N_{11} - 0.0122 \times N_{22}$ $N_{44} = -0.1470 \times N_{43} - 2.95 \times N_{40}$
V	$N_{18} = 0.0244 \times N_{16} + 0.0244 \times N_{17} + 0.8187 \times N_{19}$ $N_{24} = N_{23} + 0.8465 \times N_{25}$
VI	$N_{20} = 0.0906 \times N_{18} + 0.0906 \times N_{19}$
VII	$N_{33} = 2.35 \times N_{20} + 5.1 \times N_{23} + 1.81 \times N_{29}$
VIII	$N_{34} = N_{33} + 0.7165 \times N_{35}$
IX	$N_{36} = -0.4724 \times N_{34} + 3.3333 \times N_{33} + 2.35 \times N_4 + 2.25 \times N_9$ $+ 0.1 \times N_{13} - 0.4724 \times N_{35}$

State Nodes are : { 3,6,8,11,13,15,17,19,22,25,28,31,35,39,42 } ;   Input Nodes are : { 1, 10, 26, 37 }

TABLE 3.1 VAP control law: FSVD node computation precedence.

Parallel algorithm strategies Chapter 3

A transputer-based multiprocessor system has been programmed to compute the different node values, and the description specified in Table 3.1 has been realized for achieving good performance. A sample task assignment for the computation of the nodes of the controller case study is shown in Table 3.2.

Stage / Processor	Assignment of Tasks (N_i) to Processors (P_i)								
	1	2	3	4	5	6	7	8	9
P_1	N_{11}	N_{21}	N_{14}	N_4	N_7	N_{20}	N_{33}	N_{34}	N_{36}
P_2	N_{40}	N_{41}	N_{43}	N_{16}	N_{18}	N_{27}	N_2	N_{12}	
P_3	N_{29}	N_{30}	N_{32}	N_{23}	N_9	N_{38}	N_{44}	N_{24}	

TABLE 3.2 A 3-processors task assignment for the FSVD of the VAP control law.

3.4.6 Mapping Strategies Discretisation Methodology

The method used to discretise the continuous-time models in each of the partitioning approaches discussed in Sections 3.4.1 - 3.4.5 was the pole-zero mapping method, because it gave an acceptable approximation, and required simpler algebra, to determine discrete-time transfer functions, than other well known methods [50],[51]. The method consists of a set of rules for locating the poles, zeros and to set the gain of a z-transform, which will generate a discrete-time equivalent that approximates the given Laplace transfer function.

In the Parallel Branches and State-Space Parallel approaches, this method has been used in a modified version for computing the discrete equivalent function adding one delay in response to the unit step input. Since the resulting discrete-time transfer function has one less power of "z" in the numerator than in the denominator, the execution of the output only requires input from the previous sample time. The purpose of this was to allow one sample period in which to perform the calculation.

Parallel algorithm strategies   *Chapter 3*

In the Heuristic, Hybrid and FSVD strategies, the modified pole-zero mapping method was only used to discretise the transfer function blocks that are dependent on the control law inputs. Subsequent blocks were discretised using the original version of this method, which operates on both past and current input samples.

3.4.7 Transputer Implementation

The performance of these mapping strategies was evaluated in real-time through execution on a variety of transputer arrays. A sample implementation, using a 6-transputer array plus a Monitor transputer, and their associated occam structure is illustrated in Figure 3.10.

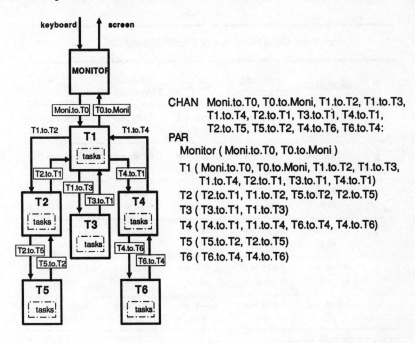

Figure 3.10 Transputer implementation and Occam structure for VAP control law.

In this structure, the Master processor, in a tree configuration, inputs data from the Monitor processor and then sends this data to the network. Next, the Master processor executes some

Parallel algorithm strategies Chapter 3

part of the tasks (a number of discrete equations) while the other processors compute the remaining tasks in parallel. The partial results are then transmitted to the Master processor which formulates the final control signals. These processes are repeated over each sample interval. The Monitor process has also been used to handle I/O devices and filing.

3.4.8 Execution Time Measurements

Execution times for the approaches described, using a varying number of transputers, are shown in Table 3.3. These times were obtained by means of using a high priority real-time transputer clock, running in the Master processor, which provides a 1 µs resolution timer.

VAP CONTROL LAW PERFORMANCE

P	HEURISTIC	PARALLEL BRANCHES	HYBRID	PARALLEL STATE-SPACE	FSVD
1	1.055 ms 100 %	2.500 ms 100 %	1.270 ms 100 %	1.071 ms 100 %	1.385 ms 100 %
2	0.760 ms 69 %	1.280 ms 98 %	0.865 ms 73 %	0.780 ms 69 %	1.050 ms 66 %
3	0.540 ms 65 %	0.910 ms 92 %	0.685 ms 62 %	0.680 ms 53 %	0.825 ms 56 %
4	0.540 ms 49 %	0.710 ms 88 %	0.565 ms 56 %	0.665 ms 40 %	0.825 ms 42 %
5	0.540 ms 39 %	0.605 ms 83 %	0.525 ms 48 %	0.655 ms 33 %	0.825 ms 34 %
6	0.540 ms 33 %	0.560 ms 74 %	0.525 ms 40 %	0.652 ms 27 %	0.825 ms 28 %

{EXECUTION TIME} [ms]
{SPEEDUP EFFICIENCY} [%]

Table 3.3 Execution Time and Efficiency of Mapping Approaches.

Efficiency, e, is computed as the speedup (elapsed time required to execute the algorithm on 1 processor, T[1], divided by the time required on p processors, T[p], divided by the number of processors, p :

$$e = \frac{T[1]}{p \times T[p]} \qquad (3.6)$$

These measurements were carried out for each of the five mapping approaches for a varying numbers of processors in the network. The programs were written in occam and run on T4-15 transputers in conjunction with an IBM-PC based transputer development system.

3.4.9 Performance Results

The performance of mapping strategies is compared in Figure 3.11. For each approach the execution time decreases with the number of Transputers in the network. As we have seen, the **Heuristic** mapping decomposes the control law into 11 tasks (Figure 3.3). These tasks are of unequal size. The three processor implementation executes the algorithm in 0.540 ms, having an efficiency of 65 %. The addition of further processors is futile since the execution of the tasks in some of the processors cannot be improved.

A similar situation is exhibited in the **FSVD** implementation, for the case studied. Here the elapsed time is dominated by the time spent for executing the critical path, computed in processor 1. The critical tasks executed in processor 1 and the increased amount of communications and synchronisation required, when additional transputers were added, made this approach efficient only for a reduced number of processors (3 in the case presented). The regularity of the matrix-vector computations expressed in the **FSVD** and the necessity of a synchronized mode of operation of the processors computing different nodes suggests that some SIMD parallel processing architecture (e.g. a systolic array) can be a more suitable.

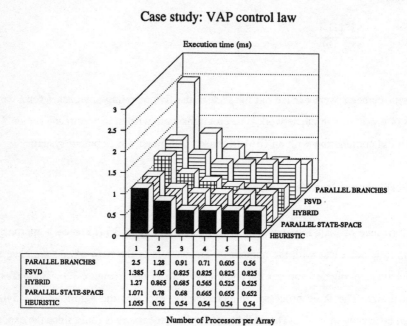

Figure 3.11 Performance of VAP control law mapping approaches.

Clearly, in the one-processor case, the duplication of effort involved in the **Parallel Branches** mapping shows up to its great disadvantage. Execution time for this approach falls more rapidly, however, though it should be noted that three processors are required to improve upon the single processor time for the **Heuristic** strategy. Using this mapping scheme, additional processors consistently reduce execution time, albeit at the expense of lower processor activity. By using six transputers, an execution time of 0.530 ms is realised using the **Parallel Branches** approach - a slight improvement on the Heuristic approach but at the expense of three extra transputers.

The **Parallel State-Space** approach, having a number of task smaller than the **Parallel Branches** approach, performed better than that approach, and just slightly slower than the **Heuristic** method, using up to 4 transputers. Further parallelism was extracted from this approach using more than 4 processors (a maximum of six transputers was used for implementing this strategy), but the communications overhead due to data routing for processors 5 and 6, at this stage, produced only a modest decrease in the execution time at the expense of lower efficiency.

The best time was obtained using to the **Hybrid** strategy on a five-transputer array. The execution time was 0.525 ms -the lowest time in the Case Study. The duplication of tasks has been limited to the critical path, $H_2(z)$, only, where parallelism has been extracted through application of the Parallel Branches approach. Hitherto, for historical reasons, the **Hybrid** method has been implemented by combining both the **Heuristic** and the **Parallel Branches** techniques. However, the **Parallel State-Space** method (a more recent and efficient strategy than the **Parallel Branches** method) can also be used to implement the **Hybrid** strategy, and evidently in a more efficient way. Nevertheless, for this Case Study, only a slight reduction in the total execution time could be foreseen when using the **Parallel State-Space** method within the **Hybrid** approach.

The **Parallel Branches** and **Parallel State-Space** methods have the virtue of minimising the communication between task groups since this mainly takes place during the initial distribution of data and the final addition of terms to form the output values. Both techniques maximise parallelism and allows more processors to be used. In contrast, the **Heuristic** and **FSVD** approaches must accept the inherent sequentialism of the original control law and the consequent limit on the number of processors that may be used. Owing to their regular approach to the decomposition of the control law the Parallel Branches and the Parallel State-

Space schemes proved to be the most straightforward to construct and, consequently, easiest to adapt, under test conditions.

The speedup and efficiency of the synchronisation and communication activities on the target multiprocessor system has determined whether or not the implementations are efficient. For this particular Case Study and some of the strategies implemented, the amount of communication and synchronisation required was large compared to the actual amount of computation each processor must carry out. However, higher levels of efficiency can be expected for implementation of more complex systems.

3.5 SUMMARY

When contemplating the implementation of control laws and associated software on parallel processing systems, it is required to organise the control law to realise the maximum benefits of parallelism and to allocate, on the target system, the resulting tasks in a convenient way. The importance of an uncomplicated mapping scheme in such a safety critical application motivated research into tools to automate the partitioning and task allocation processes. To support the application of parallelism in this way, a number of tools have been developed. These will be described and evaluated in Chapter 4 of this book. The aims of these tools are: to automate the implementation of control algorithms on a parallel processing transputer-based system and to assist the control engineer to determine the likely speedup to be achieved, the number of processors to be used, and experiment with alternative system topologies. These topologies are evaluated by displaying, on-line, task allocation, processor activity and execution time data.

CHAPTER 4

EPICAS - AN ENVIRONMENT FOR PARALLEL IMPLEMENTATION OF CONTROL ALGORITHMS AND SIMULATION

4.1 INTRODUCTION

This Chapter describes how we have integrated, the control system design package, MATLAB, with the Transputer Development System (TDS), to generate an Environment for Parallel Implementation of Control Algorithms and Simulation (**EPICAS**). This environment offers the control engineer a number of software tools for automating the implementation of control algorithms and simulation systems on transputer-based architectures. The tools are used to map systems onto transputer architectures of different sizes and topologies, and to evaluate strategies, by displaying, on-line, task allocation, processor activity and execution time data.

4.2 EPICAS

EPICAS (Enviroment for Parallel Implementation of Control Systems and Simulation) integrates the control system design package, MATLAB, with a Transputer Development

System (TDS). It has been developed originally for running on a PC environment, but a SUN-based version of this system is in development. It is also being modified to run with the Occam Toolset System. EPICAS offers the control engineer a number of software tools to automate the implementation of control algorithms on transputer-based systems. The need for such an environment arose out of our previous work on mapping strategies for implementing controllers on transputer networks. The effort involved in human-engineered implementations, described in Chapter 3, stimulated the research into the development of software tools to automate the implementation process. The tools within the environment, see Figure 4.1, are grouped in two sets: *MATLAB tools* for parallel partitioning, and *OCCAM tools* for task allocation.

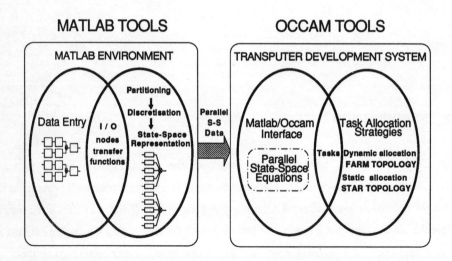

Figure 4.1 EPICAS -an environment for mapping systems on transputers.

The tools for parallel partitioning, based on MATLAB, generate a parallel representation of a system (e.g. a control law) as a number of independent tasks. A continuous-time system description in block diagram form may be input. The software performs the partitioning, discretisation and parallel representation of the system as a number of state-space equations.

The OCCAM tools for task allocation, developed in TDS, automate the mapping of the tasks (state-space equations) onto a number of transputer-based topologies, using either static or dynamic allocation strategies. These tools permit the evaluation of both types of strategies, by displaying on-line task allocation, processor activity and execution times on the target hardware

4.3 MATLAB TOOLS FOR PARALLEL PARTITIONING

Efficient parallel implementation of control systems depends on a suitable partitioning process. Here we describe software tools based on MATLAB which were developed and used to automate the parallel partitioning of control algorithms into modules or tasks for further scheduling on a parallel processing transputer-based system.

4.3.1 MATLAB Overview

MATLAB (**Matrix Laboratory**) [52] was first released in 1980 as a collection of software tools of linear algebra, matrix computation and numerical analysis to assist scientists and engineers with their numerical calculations. Complex arithmetic, eigenvalues, root-finding, matrix inversion, and FFT's are examples of these tools. Elementary matrix commands for data analysis and statistical tools are also provided. The original version of MATLAB was written in FORTRAN. A second generation of MATLAB, written in C language, was designed as an integrated system, including graphics, programmable macros, IEEE arithmetic, a fast interpreter and many analytical commands. On PC- compatible MATLAB is called PC-MATLAB [8]. On larger computers, like SUN Workstations and VAX computer, the new version of MATLAB is known as PRO-MATLAB [53]. MATLAB is now available for parallel processing computers such as Convex and Alliant [54].

MATLAB essentially provides the user with a set of low-level commands and the prospect of defining new commands by enabling the user to create M-files (files containing a sequence of MATLAB statements distinguished by having the extension ".m"). M-files are a means of automating long sequences of commands. The new files are subsequently added to the set of available commands. MATLAB capabilities have been extended by the provision of Toolboxes (a collection of M-files with specific aims).

A number of toolboxes, using MATLAB matrix functions, offers specialist control engineering software for common control system design, analysis, and modelling techniques. Through these extensions the following toolboxes are now available: Control System [55], Robust-Control [56], Multivariable Frequency-Domain [57], Signal Processing [58], System Identification [59] and State-Space Identification [60]. MATLAB has become a de facto standard for control systems research and design, and new tools are expected to be developed in a number of applications.

4.3.2 Parallel Partitioning MATLAB Toolbox

The MATLAB tools, developed for partitioning of control algorithms, are a collection of customised M-files. This Toolbox includes utilities for data entry, block partitioning, and parallel representation, which are called *blkentry.m*, *blkpart.m* and *paradata.m* respectively, see Figure 4.2. The toolset enables the user to perform an automatic partitioning of the control algorithms into a number of independent tasks.

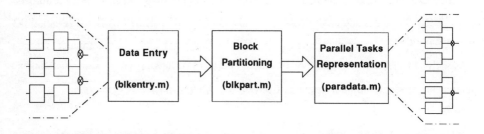

Figure 4.2 MATLAB tools for parallel partitioning of systems.

In this study, the RAE Versatile Autopilot Control law (VAP), described in Chapter 3, is used to illustrate the application of the tools for a specific flight control case. As this controller was chosen as a test algorithm for development purposes, the discussion is focused on this control law. However the Toolbox described here is, of course, valid for a variety of other dynamic systems which can be modelled as a collection of interconnected blocks.

4.3.2.1 Data Entry Process

As practical controllers are often described in block form, block diagram entry facilities are provided to accept this type of system description, in order to generate a more convenient representation such as the state-space model. The new MATLAB tool, *blkentry.m*, has been generated to provide a means of inputting system information into the package. An overview of this block diagram entry process is shown in Figure 4.3.

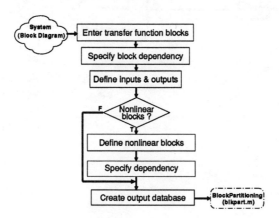

Figure 4.3 Block diagram data entry process.

This tool facilitates the entry of information arising out of a block diagram description of the system. Figure 3.1 shows a typical example of a block diagram representation for the VAP control law Case Study and is repeated here in Figure 4.4.

EPICAS   Chapter 4

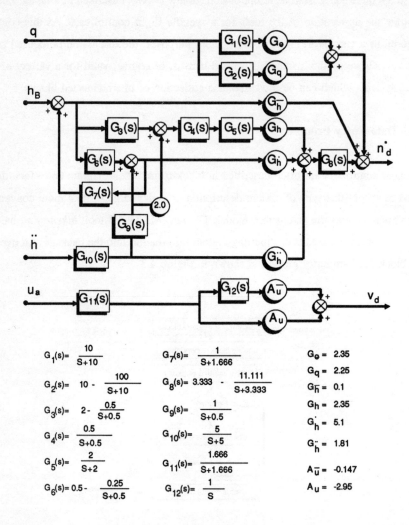

Figure 4.4   Case study: VAP Control Law.

The input process accepts the coefficients of the numerators and denominators of the transfer function blocks. Block connectivity and system inputs/outputs must also be specified. Systems containing nonlinear elements, cascaded to the controller outputs, can also be input. These elements may be highest-wins selectors, lowest-wins selectors, and relays blocks. Figure 4.5 shows an example of a simplified controller containing two types of these nonlinearities (a highest wins block that chooses the highest signal from a number of input signals, and a relay block for selecting an input, controlled by a selection flag).

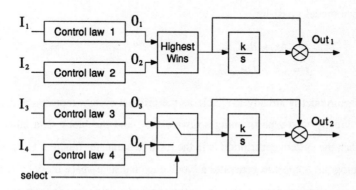

Figure 4.5 A simplified controller with nonlinearities.

In addition, as proportional plus integral (P+I) terms are commonly cascaded to the output stage of typical flight controllers, these terms are also considered and can be input during the data entry session. This additional procedure then includes the input of the type of nonlinear block (lowest wins, highest wins, relays, etc.), the blocks connected to their inputs and the P+I block data.

When a state-space description of the system is already available as the initial system representation, the block diagram data entry tool will not be necessary and the state-space model may be input directly. However, for systems of even moderate complexity, it can be quite difficult to find the state-space model required without computer aids. Using system block diagrams and a number of dedicated MATLAB functions, state-space models can be automatically generated; this process is described in Section 4.3.2.2.

4.3.2.2 Partitioning Process

The MATLAB tool, *blkpart.m*, transforms the control law into a number of independent tasks, by means of obtaining the relationship between each input and output variable, reducing the control algorithm to a set of independent path transfer functions. The utility accepts the block description produced by the data entry tool outlined in Section 4.3.2.1, in order to generate the state-space model given by

$$\dot{x} = Ax + Bu \qquad (4.1)$$
$$y = Cx + Du \qquad (4.2)$$

If a state-space model is already available as the original system representation, this model is supplied directly to the partitioning software tool, and the block data entry tool is not required. When the system description is in the form of a transfer function block diagram, see Figure 4.6, program *blkpart.m* generates a block diagonal state-space model consisting of the unconnected transfer functions. Using the interconnection data, the program connects up the block diagonal model and returns a state-space model.

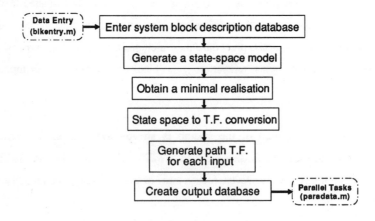

Figure 4.6 Overview of partitioning process.

Redundant or unnecessary states, introduced by constant gain blocks in the block diagram, are removed by transforming the matrices A,B and C to controllability and observability staircase forms using available MATLAB functions. This transformation isolates the uncontrollable and unobservable states, which are then removed from the model. The relationship between each input and output variable is obtained by applying the state-space to transfer function conversion called **ss2tf.m**, for each system input. This process calculates the transfer function

$$H(s) = \frac{NUM(s)}{den(s)} = C\,(sI-A)^{-1}\,B + D \qquad (4.3)$$

of the system represented by Eqs. (4.1) and (4.2). Vector **den** contains the coefficients of the denominator in descending powers of **s**. The numerator coefficients are returned in a matrix **NUM** with as many rows as there are outputs. Each nonzero row of coefficients in the matrix **NUM** is associated to the common denominator **den** to form a path transfer function. The function **ss2tf.m** is applied as many times as there are inputs. Figure 4.7 illustrates the resulting data of *blkpart.m* for the VAP flight controller Case Study.

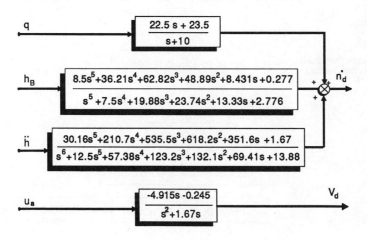

Figure 4.7 VAP control law: Path transfer functions.

4.3.2.3 Generation of Parallel Tasks Representation

The path transfer functions generated during the partitioning process described in Section 4.3.2.2, are frequently higher order functions. In realising the digital implementation of these models important sources of error are those due to round-off in arithmetic operations, quantisation of input signal and, quantisation of the coefficients a_i and b_i [61]. These errors arise because of the practical limitations of the number of bits that represent the signal samples and coefficients. Errors associated with quantisation of the coefficients become large as the order of the transfer function is increased. In a high-order model (experience suggests order >= 5) implemented in direct programming or in standard programming, small errors in the coefficients cause large errors in the locations of poles and zeros in the model. This type of error, however, may be reduced by mathematically decomposing a higher-order transfer function into a combination of lower order models to produce a system less sensitive to coefficient inaccuracies. This formulation also extracts more parallelism from the models supplied by the partitioning tool.

The procedure to generate the final parallel representation of a system, is automated by a specially developed function called *paradata.m*, and is described in Figure 4.8.

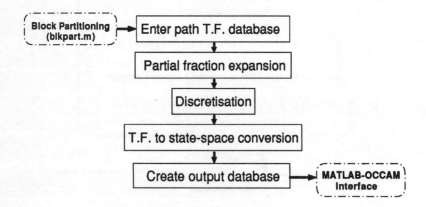

Figure 4.8 Parallel tasks representation process.

The MATLAB tool accepts the path transfer functions provided by the partitioning utility *blkpart.m*, and decomposes each model into a parallel connection structure of lower-order partial fractions (first and second order). The resulting functions are discretised using the pole-zero mapping method and finally represented as a parallel connection of discrete-time state-space equations. Figure 4.9 shows the VAP control law as a parallel connection of state space equations, for T = 0.1 s. Also note that x_1, x_2, ... , x_{14} represent vectors of dimension 1 or 2. This tool also produces an output database, which includes the coefficients of the state-space equations and their association to form the controller outputs. The output database is ported to the Transputer Development System, where a set of OCCAM-based tools has been written, to map this partitioned representation of a system, using alternative strategies of tasks allocation, on a parallel processing transputer-based network.

4.3.2.4 Discretisation

The method used to discretise the continuous-time models was the pole-zero mapping method, already mentioned in Chapter 3. The models were discretised using dedicated MATLAB functions (**c2d_pz.m** and **c2d_mpz.m**) provided by a CAD toolset developed in-house [49]. These functions implement the pole-zero mapping method and its extended version respectively. Alternative discrete equivalents can be used (bilinear, backward, rectangular, etc). These are provided in a currently available Digital Control for Dynamic Systems MATLAB Toolbox [62].

4.3.3 MATLAB Tools Remarks

The aim of the partitioning toolset has been to provide a means of automatic parallel partitioning of control algorithms. We have not provided, however, a graphics environment for entering the system description, because this facility is already available in other commercial packages such as Simulab [63], and Protoblock [64], where systems are defined much like drawing a block diagram, choosing individual blocks from standard block libraries supplied or from user defined libraries.

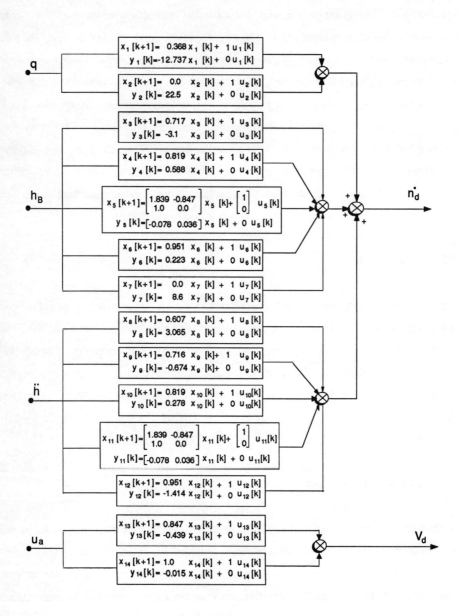

Figure 4.9 VAP control law: Parallel tasks representation.

A critical aspect common to the software tools, and in general to the MATLAB environment, is the run-time efficiency of the user-defined functions. As the command language provided by MATLAB is an interpretive environment, the user-defined M-files run slower than the built-in algorithms programmed directly into the package implementation language. It would be therefore very useful to have facilities to translate user defined M-files from the command language, once they are working properly, into built-in functions on the implementation language. As the partitioning Toolbox is based on the basic capabilities of MATLAB and its associated Control System Toolbox, these utilities have an intrinsic low run-time performance. This aspect, however, is not critical because the partitioning process takes place off-line.

4.4 OCCAM TOOLS FOR TASK ALLOCATION

A set of tools for allocation of tasks onto transputer-based systems has been developed using the Transputer Development System. The toolset, depicted in Figure 4.10, has been written in the occam 2 language and is used to automate the mapping of control algorithms onto a number of transputer topologies, using both static and dynamic task allocation strategies.

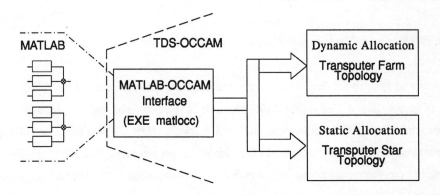

Figure 4.10 OCCAM tools for tasks allocation.

A control algorithm, partitioned by the MATLAB tools into a set of independent state-space equations, can be imported into the TDS, using a ***MATLAB-OCCAM interface*** tool, and mapped onto a parallel processing transputer-based network. The mapping of the control algorithm is achieved using tools that implement either dynamic or static task allocation strategies. The dynamic allocation tools are based on a ***Processor Farm*** computational model. The static allocation toolset is based on a ***Processor Star*** topology. The toolset permits the evaluation of both approaches, by displaying processor activity and performance.

In the following sections, the partitioned version, of the RAE Versatile Autopilot Control law (VAP), will be used to illustrate the application of the tasks allocation tools to a working example of digital flight control. In order to map the control algorithms considered onto a transputer network, a purpose-built Eurocard transputer system [65], developed in-house for real-time control applications, has been used. This system includes a number of T414/T800 single transputer boards as basic building blocks, accommodated in a Eurorack, along with a variety of support boards. These include a crossbar switch, system reset/analyse, ADC/DAC and digital I/O cards. This modular system is portable and customisable. It allows the user to select only the modules of direct use, in order to distribute the limited transputer resources as efficiently as possible.

4.4.1 Transputer Development System Overview

A number of systems for developing occam programs for transputers is available. The most typical is the Inmos IMS B004 [66] which is a plug-in card for the IBM PC (or compatible). This card, together with the TDS software (IMS D700D) [67], provides an environment for the development and execution of occam programs. The TDS includes an interactive programming environment which has a **folding** editor. This provides the ability to hide blocks of lines (text or data) by storing them as fold lines within the document. Fold lines are represented on screen by three fullstops (**... fold line**). Data can also be stored in separate files, as filed folds in the filing system. A variety of libraries of mathematical functions, I/O and debugging facilities are also provided. A server program interfaces the TDS to the PC terminal and filing system. A typical example of this arrangement is shown in Figure 4.11.

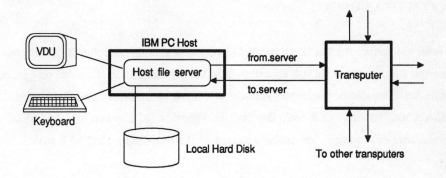

Figure 4.11 The transputer / host development relationship.

Using the TDS environment the user can edit, compile, and run an OCCAM program to execute on

- the TDS board transputer,
- a network of transputers connected to the TDS board, or
- on a standalone transputer system, independent of the TDS.

Before compilation, the program fold must be filed, and enclosed within a compilation fold (*foldset*). The main types of foldsets are

- **EXE** - for a program which executes within the TDS.
- **PROGRAM** -for a program which executes on a network of transputers.

For SUN systems, an occam 2 Toolset [68], is available. This provides a platform for development of mixed language programs for transputers. In this context we will refer to "host computer" as the combination of transputer board and the development platform (PC or SUN), and the transputer network will be referred to as the "target system".

4.4.2 MATLAB-OCCAM Interface

In order to interface the MATLAB parallel representation of the controller (as a collection of state-space equations) to the task allocation OCCAM software, for the final mapping process, within the transputer development system, a MATLAB to OCCAM link utility called *EXE matlocc* was written. This tool, depicted in Figure 4.12, imports into the transputer development environment, the database created by the partitioning MATLAB tools.

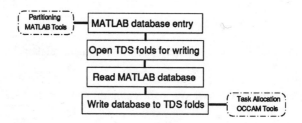

Figure 4.12 Overview of MATLAB-OCCAM interface process.

The database is converted to TDS format and stored as a number of structured files. These new files are used as an input database for a set of OCCAM utilities that performs the mapping of the parallel equivalent version of the controller on a transputer-based system.

4.4.3 Dynamic Task Allocation Tools

The dynamic allocation tools are a number of OCCAM programs based on a *Processor Farm* computational structure. Processor farming is a conceptually simple method of utilising multiple processors. The operation of a processor farm is based around the ability to dynamically schedule tasks onto a collection of processing elements. The scheduling action must occur very frequently, and as a worse case, will equal the number of tasks in the partition. It must be simple, so that it does not present too large an overhead.

4.4.3.1 Processor Farm Linear Topology

The most typical version of the processor farm model uses a single *master* processor, for scheduling tasks, to one or more *worker* processors connected in a line. This model is shown in Figure 4.13.

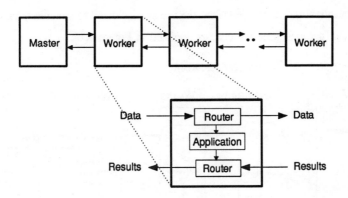

Figure 4.13 Processor farm model.

The application code or computational task (in our case, a first- or second-order state-space model, plus all the coefficients of the partitioned control algorithm to form the state-space equations) runs on each of the *worker* processors and routers are used to route data and results through the network. Task identifiers, to define which tasks must be executed, and input data are passed to the network as required, in order to utilise the full processing potential of the system. This provides an automatic mechanism for dynamic load balancing. Results are collected by the *master* as they are calculated by the workers. Finally, these results are used to compute the system outputs.

4.4.3.2 Transputer Implementation of the Processor Farm

This implementation of the farm model was modelled by a set of occam processes running in parallel. A *master* process, mapped onto a single transputer, controls the organisation and execution of tasks on its subordinate *worker* processes, allocated onto their own transputers. The workers are modelled as identical parallel occam processes. When each process completes the given task, further work is distributed for completion until the whole problem is complete. Ideally each *worker* process may be connected via a channel to the *master*. Since the transputer has only four links, the workers must be distributed over several lines of transputers. In this initial implementation, the processor farm was structured as a single line of workers, with the master at one end, as depicted in Figure 4.14.

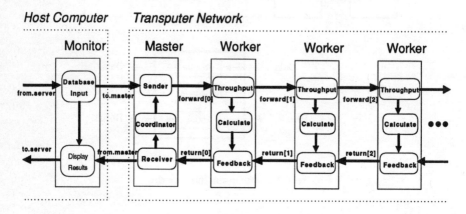

Figure 4.14 Transputer implementation of a linear processor farm.

Then, each *worker* process is not just concerned with the execution of its task. It also acts as a message router, passing on data to processes further down the farm. An additional process called *monitor*, running on the host computer, was developed to read the partitioned controller database generated by the MATLAB-OCCAM program interface, and send it down the farm. The database is used by the workers to identify the coefficients of each allocated state space equation for the task to be executed. Process *monitor* also receives the farm output database and displays the processor activity and performance of the system. The general form of this farm in terms of occam is given by

>... global declarations
>**PAR**
> ... **master** process
> **PAR** index = 1 **FOR** number.of.workers
>   ... **worker** process

where *global declarations* includes a constant defined for the number of workers in the transputer farm, which is all that needs to be changed to use a different number of transputers. Link declarations and channels are also placed in this fold.

A *worker* process comprises three processes: *throughput*, *calculate* and *feedback*. This array uses a **PRI PAR** construction to implement top priority to data and results routing and thus guarantee a high throughput for communications. This is important for the processes which use the transputer links, so that messages are transmitted without delay. If a high priority process was not used, the message would not be examined until the message switch was scheduled by the low priority round-robin scheduler of the transputer [69]. Such a structure is depicted in Figure 4.15 and it is expressed in occam as follows

> **PROC** worker
>   **PRI PAR**
>   **PAR**
>   ... throughput
>   ... feedback
>   ... calculate
>   :

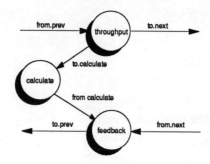

Figure 4.15 The component processes of a worker process.

The *throughput* process is responsible for accepting data packets (consisting of a processor identifier, task identifier and input data) from the *master* process, sending data for its *calculate* process, and forwarding excess data to workers further down the farm. Data is passed from the *throughput* process each time the *calculate* process completes the previous calculation (the evaluation of a first order or second order state-space equation) and requires more work. Then, the resulting data is passed on to the *feedback* process. The *feedback* process multiplexes the results from its *calculate* process and those received from other workers on the farm, and feeds them back to the *master* process. This process is conveniently implemented using an **ALT** construction. Representing these processes in occam gives

```
-- throughput
WHILE tasks.are.available
  SEQ
    ... receive data packet
    IF
      worker.number = my.number
        ... send data to calculate process
      TRUE
        ... forward data packet to next worker
```

-- **calculate**
WHILE tasks.are.available
  SEQ
    ... receive data from throughput process
    ... evaluate state space equation
    ... send results to feedback process

-- **feedback**
WHILE tasks.are.available
  ALT
    ... receive results from calculate & send to previous worker
    ... receive results from other workers & send to previous

On the other side, the *master* interfaces to the host computer to receive the controller parallel description database plus controller inputs data. It generates and sends data packets associated with this description (processor and task identifiers plus input data) to the workers in the farm. The *master* process comprises three processes running in parallel: *sender*, *coordinator* and *receiver* (see Figure 4.16).

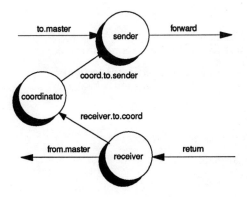

Figure 4.16 The component processes of the master process.

The occam structure of the *master* process is

    PROC master
      PAR
        ... sender
        ... coordinator
        ... receiver
    :

The *sender* process accepts inputs from the *monitor* process, and processor and task identifiers from the *coordinator* process, forwarding this data down the farm. The *coordinator* process initialises the farm by issuing **number.of.workers** data packets to the work force. This data just fills up each worker. As each worker is reported processed, another task is issued to the farm, until all the equations have been evaluated. At this point a termination notice is issued to the farm. Finally, the *receiver* process is responsible for accepting the results generated by the farm, plus identifiers, sending back the identifiers to the *coordinator* process, which passes the now free worker number to *sender*. The occam representation of these processes is

  -- **sender**
  ... receive inputs from monitor process
  WHILE tasks.are.available
    SEQ
      ... receive worker identifiers from coordinator process
      ... send worker, task identifiers & inputs to farm

  -- **coordinator**
  ... issue initial tasks to workers via sender
  WHILE tasks.are.available
    SEQ
      ... collect worker-free identifier from receiver process
      ... send new tasks to worker-free via sender

```
-- receiver
WHILE tasks.are.available
    SEQ
        ... receive worker & task identifiers from farm plus results
        ... send worker-free identifier to coordinator process
        ... route results & worker activity to monitor process
```

The final occam implementation of a processor farm will comprise a PAR constructor containing instances of the *master* process and a number of replicated *worker* processes, as follows

```
PAR
    master (forward[0], return[0])
    PAR i = 1 FOR number.of.workers
        worker (forward[i-1], forward[i], return[i-1], return[i])
```

The processor farm model described in this Section was implemented in an occam **PROGRAM** called *farmline* and used to allocate dynamically the partitioned version of the VAP control law (represented as 14 independent discrete-time equations). The user can select and modify, at run-time, the number of *active* workers to be used in the farm, from a maximum number of workers, previously set at compilation-time. The utility output provides for visualisation of the *performance* of the array, by presenting figures of *execution time* and *processor activity*.

Figure 4.17, displays the architectural arrangement and the task activity of each processor for a farm (using from 1 up to 6 workers). The data below each worker represents the number of each task (state-space equation) processed and its associated order (first or second order), for tracing the history of the computation within each iteration time. The execution time of the linear arrangement of the farm was measured using the real-time clock facility of the transputers. This has been determined by calculating the differences of *timer* input statements

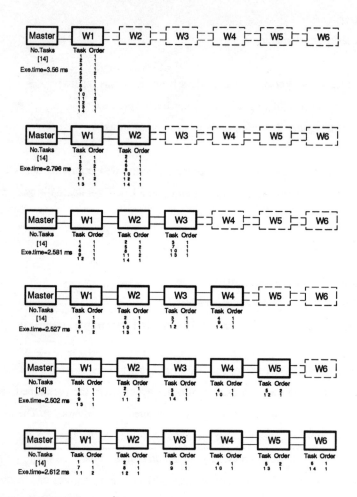

Figure 4.17 Processor activity: farm using from 1 up to 6 workers.

placed within the master process, because this process coordinates and synchronises the rest of the processes. The measurements were carried out varying the number of workers in the farm, from 1 up to 6 workers, in the test case presented. Figure 4.18 summarises the performance of this array, considering only the elapsed time or execution time for the two types of transputers (T414-15 and T800-20) available in the rack system used (other important metrics will be considered in Chapter 5). The *execution time*, in each case, is the

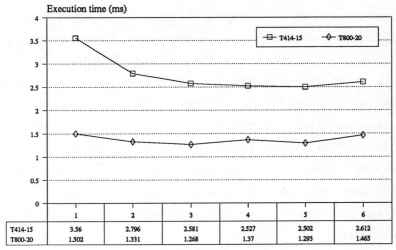

Figure 4.18 Performance of a processor farm -linear array.

interval in which the master executes a complete iteration (i.e. allocates tasks to the farm of workers, receives results of each task from the farm and computes the global system outputs). Figure 4.18 presents results corresponding to the elapsed time to run the same particular job (the VAP control law) on a processor farm with a varying number of workers. Using T414 transputers, a modest but significant reduction in the execution time was achieved as additional processing power is introduced. However, no extra improvement could be attained for more than 5 workers. Furthermore, for 6 workers or more the execution time was higher than the previous cases. Here, the communication overhead is the major element that affects the performance of the farm and is exacerbated when the linear array is long. The use of the T800 transputer obviously showed a superior reduction in the elapsed time, when compared with the T414 (note that the typical T800 floating point operations, used in this application example for the calculation processes, are about 500 ns [70], whereas the operations in the T414 transputer have a average value of 10µs).

Further research into the development of the techniques of process farming has been accomplished to improve the original implementation of the processor farm. These techniques include the use of task buffering within each worker, so that a new task can be executed with minimal latency when a worker becomes idle. Another approach comprises an extended model of the processor farm using three lines of workers arranged around the master, thus reducing the messages throughput and increase the bandwidth of data through the master. These improvements will be described and their results analysed in Chapter 5. This analysis will consider other standard metrics, apart from *execution time*, such as *speedup, speedup efficiency* (commonly used for measuring the performance of parallel processing systems), and a novel metric called *serial fraction*, which provides additional information, not revealed, by other used metrics.

4.4.4 Static Task Allocation Tools

Tools for static task allocation has been also developed during this research. This toolset is based on a *Processor Star* model as computational structure. This model employs a central process for distributing a package of fixed tasks to a number of worker processes arranged around the central node. The tasks are allocated to the workers and bound them to them for their lifetime. Each worker executes a fragment of the application code (in our case, a sub-set of the state-space equations of the model). Communications between task groups takes place during the initial distribution of data and the final computation to form the output values.

4.4.4.1 Processor Star Topology

This processor star approach has been modelled by a set of processes running in parallel. The *master* process sets the organisation a number of fixed tasks to be executed by its satellite *worker* processes, see Figure 4.19.

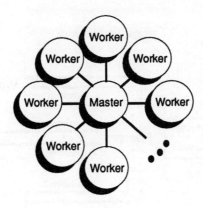

Figure 4.19 Processor star topology.

The application code resides in the workers. The master sends an identifier to each available worker to state which component of the application must be executed. The master also executes a portion of the computational task. When the workers complete their job, these send the partial results back to the master, which in turn calculates a final computation.

4.4.4.2 Transputer Implementation of a Processor Star

A transputer implementation of the processor star topology, displayed in Figure 4.20, shows a *master* and three *workers*. Three workers were used because of the limited number of transputer links. However, the use of a cross-bar link switch such as the INMOS IMS C004 [71], would increase the number of workers employed. The fourth link of the central processor is interfaced to a *monitor* process into the Transputer Development System, as in the processor farm case, for I/O and file handling facilities. The occam program associated with this implementation is compiled for three workers, but at run-time can select a smaller topology according to the number of active workers.

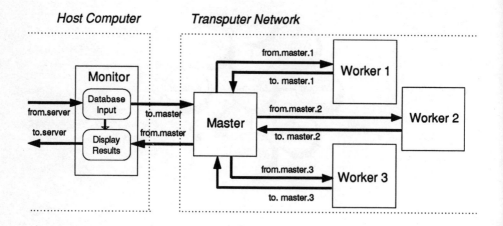

Figure 4.20 Transputer implementation of a processor star array.

The occam structure for this arrangement can be written as follows:

PAR

    ... **master** process

    ... **worker1** process

    ... **worker2** process

    ... **worker3** process .

These processes (in a simplified occam description) can be outlined as:

-- **master**
WHILE TRUE
  SEQ

... set topology (0,1,2,3 workers), send topology to workers
CASE topology
  0
    ... perform all the tasks (N tasks)
  1
    ... perform N/2 & remainder tasks, get results from Worker 1
  2
    ... perform N/3 & remainder tasks, get results from Workers 1 & 2
  3
    ... perform N/4 & remainder tasks, get results from Workers 1, 2 & 3
... perform supplementary calculations & send to monitor

-- worker1
WHILE TRUE
  SEQ
    ... receive topology identifier
    CASE topology
      1
        ... execute N/2 tasks, send tasks results to master
      2
        ... execute N/3 tasks, send tasks results to master
      3
        ... execute N/4 tasks, send tasks results to master
    ELSE
      SKIP

-- worker2
WHILE TRUE
  SEQ
    ... receive topology identifier

```
        CASE topology
           2
              ... execute N/3 tasks, send tasks results to master
           3
              ... execute N/4 tasks, send tasks results to master
           ELSE
              SKIP

-- worker3
WHILE TRUE
   SEQ
      ... receive topology identifier
      CASE topology
         3
            ... execute N/4 tasks, send tasks results to master
         ELSE
            SKIP
```

The processor star model described in this Section was implemented in an occam **PROGRAM** called *procstar*. In order to apply this model to a working example, the partitioned version of the VAP control law Case Study was used. The user can select and modify, at run-time, the number of *active* workers to be used, from a maximum number of workers previously set in compilation-time (three workers in this case). The array performance can be visualised by presenting figures of *execution time* and *processor activity*. Figure 4.21 displays the architectural arrangement and the task activity of each processor for different topologies, using up to 3 workers. The data below each worker represents the number of each task processed and its associated order. Occam *timers*, within the master process, were used to measure the performance of the star topology. Measurements were carried out by varying the number of workers in the array, from 0 (when the master performs the whole algorithm) up to 3 workers, in the test case presented.

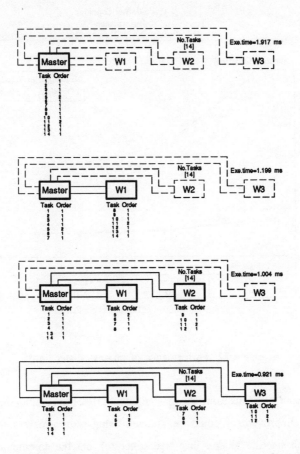

Figure 4.21 Activity map for a star array using up to 3 worker.

Figure 4.22 shows the performance of this array, using T414-15 and T800-20 transputers. Here, *total time*, is defined as the interval in which the master executes a complete iteration (performing its own tasks, receiving task results from the workers and performing the supplementary calculation to generate the global system outputs).These results correspond to the elapsed time to run a specific job (the VAP control law) on a processor star array with a varying number of satellite workers. The T414 transputer implementation displays a very significant reduction in the execution time, when additional processing power was introduced

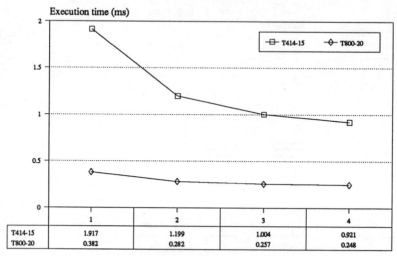

Figure 4.22 Performance of processor star array.

into the system: from 1.917 ms using the master only, to 0.921 ms employing three additional workers. Further improvement could be attained using more workers (connecting extra processors to each satellite worker in a tree structure), but the communication overheads introduced by routing information, would make the improvement very modest. At the same time, at this point (using 4 processors) the latent parallelism in the algorithm seems to be exhausted. A better alternative was the use of T800 transputers. This evidently exhibited a considerable improvement in the execution times, when compared with the T414 implementation: from 0.382 ms with the master only, to 0.248 ms using 3 workers. In fact, it is interesting to note that a single T800 transputer can perform the computation of this Case Study, about 2.5 times faster than the array of four T414 transputers. An extended analysis of the performance of these tools of static task allocation will be described in Chapter 5.

4.4.5 OCCAM Tools Remarks

The Occam tools developed permit the automatic mapping of control algorithms onto a number of topologies. This toolset enables evaluation of both static and dynamic task allocation strategies, by displaying processor activity and performance. An important point to note here is that the different strategies are tested on the **actual** hardware, thus avoiding simulation inaccuracies associated with imperfect knowledge of system behaviour (e.g. task switching, communications). In a Case Study, the VAP flight controller was transformed into a set of tasks by using the partitioning MATLAB tools and was employed to illustrate and evaluate the application of these allocation tools in a specific flight controller example. In its basic realisation a processor farm (using dynamic task allocation) displayed an inferior performance when compared with the static approach. Communications overhead is a significant component in the execution time and is exacerbated when the linear array is long. The static allocation approach exhibited a better performance. This strategy, has the virtue of minimising communication between task groups since this only takes place during the initial distribution of data and final computation to form the output values. Performance measuring of dynamic and static task allocation tools has been outlined in Sections 4.4.3 and 4.4.4 , for the VAP control law case study. However the results of this analysis, evidently, can vary very strongly with the algorithm and size of different problems. This consideration led us to apply the software tools developed to more complex problems, in order to analyse the effects, on the tools performance, when the problem complexity is increased. These aspects will be also investigated in Chapter 5.

4.5 USING EPICAS FOR SIMULATION

The MATLAB and OCCAM Toolboxes outlined in Sections 4.3 and 4.4 have been depicted using a particular flight control law, chosen as a test algorithm for development purposes. Although the Toolboxes were originally developed for mapping controllers onto transputer networks, these are also eminently suitable for simulation purposes. The tools have provided a means to automate the partitioning and scheduling of dynamic systems for multiprocessors.

This is illustrated in the work developed by the authors of Reference [72], where the models of the flight dynamics of a BAC 1-11 short-haul transport aircraft, provided by RAE Bedford, have been partitioned and mapped onto a transputer-based system using these Toolboxes, in order to observe its response to given step inputs.

4.5.1 Aircraft Flight Dynamics Models

The linearised, longitudinal equations of motion of the BAC 1-11 were provided by RAE Bedford, and are given in Eqs. (4.4)-(4.6). The equations assume an approach airspeed of 140 knots in a fixed configuration with 26° flaps and undercarriage lowered.

Forces normal to flight path:

$$(\dot{\alpha}-q) = -0.1375(u_{gr}+u_w) -0.7134(\alpha+\alpha_w) -0.232q -0.010\theta -0.0467\eta \tag{4.4}$$

Forces along flight path:

$$\dot{u}_{gr} = -0.04669(u_{gr}+u_w) -0.1590(\alpha+\alpha_w) -0.3326\theta +0.0647T \tag{4.5}$$

Pitching moment:

$$\dot{q} = -0.9215(\alpha+\alpha_w) -0.680(q-\alpha_w) -1.2975\eta -0.227(\alpha+\alpha_w) \tag{4.6}$$

The input states to the aircraft model are the demands from the control law

η = elevator position (degrees)

T = thrust (kiloNewtons)

and the disturbances

u_w = horizontal gust (ft/second),

α_w = change in incidence due to vertical gust(degrees).

Relating

α_w to vertical gust v_w, then $\alpha_w = 0.2423\ v_w$.

The states generated from the equations of motion are

q = pitch rate (degrees/second), where

$$q = \int \dot{q}\, dt, \tag{4.7}$$

θ = pitch attitude (degrees), where

$$\theta = \int q\, dt, \tag{4.8}$$

u_{gr} = change in ground speed (knots),

α = change in inertial incidence (degrees).

The output states, i.e. those required by the control law, are

u = change in airspeed (TOTU) in knots, where $u = (u_{gr}+u_w)$,

q = pitch rate (QRATE) in degrees/second,

n = normal acceleration (AAZCG) in "g" units, normal to the flight path, where $n = -0.124(\dot{\alpha}-q)$,

h = change in height (HERROR) in feet. This is obtained by double integration of n

$$\dot{h} = 32.18 \int n\, dt, \tag{4.9}$$

$$h = \int \dot{h}\, dt \tag{4.10}$$

4.5.2 Simulation using TSIM

The BAC 1-11 equations of motion were first simulated using TSIM, the nonlinear dynamic simulation package [73]. A TSIM description of the aircraft flight dynamics was developed, and its behaviour tested for individual input step input excitations, in order to provide a reference for verifying the transputer-implemented models. Figure 4.23 shows the TSIM model response for a step input of 20 kiloNewtons in the thrust. Figure 4.24 displays the model response to an input of 1 degree in the elevator position.

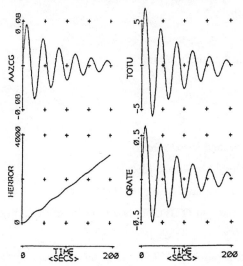

Figure 4.23 TSIM model response for step input of T = 20 Kilonewtons.

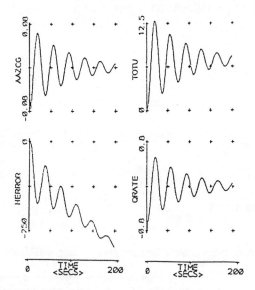

Figure 4.24 TSIM model response for step input of η = 1 degree.

4.5.3 Simulation using EPICAS

In order to simulate the models of the aircraft flight dynamics on a transputer-based system, a block diagram representation of the Eqs. (4.2)-(4.4), also provided, was used. This was a suitable means for entering the system into the EPICAS tools for partitioning. The resulting flow diagram of these models is presented in Figure 4.25.

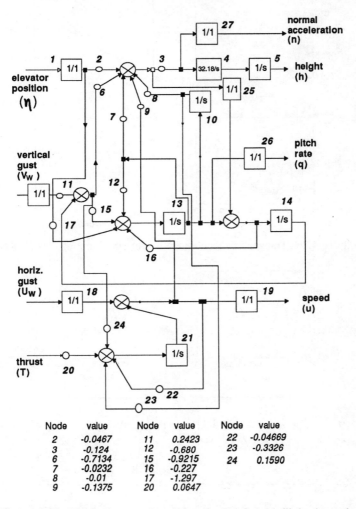

Node	value	Node	value	Node	value
2	-0.0467	11	0.2423	22	-0.04669
3	-0.124	12	-0.680	23	-0.3326
6	-0.7134	15	-0.9215	24	0.1590
7	-0.0232	16	-0.227		
8	-0.01	17	-1.297		
9	-0.1375	20	0.0647		

Figure 4.25 Block diagram describing the BAC 1-11 flight dynamics.

First, the MATLAB toolbox was applied to generate a parallel representation of the system Then the OCCAM toolset was used to automate the mapping of the parallel tasks onto one or more transputers according to a selected topology. Finally, individual step inputs were applied to the system inputs. Figures 4.26 and 4.27 show the system response to the same step inputs applied in the TSIM simulation. These results were found to agree with the TSIM model.

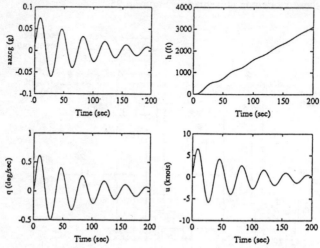

Figure 4.26 Transputer model response for step input of T = 20 KN.

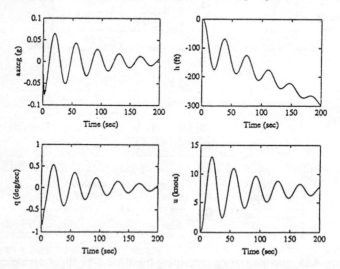

Figure 4.27 Transputer model response for step input of η = 1 degree.

4.6 SUMMARY

The integration of the control system design package, MATLAB, with a Transputer Development System (TDS), has led to the emergence of **EPICAS** (an Environment for Parallel Implementation of Control Algorithms and Simulation). This environment offers the control engineer a number of software tools for automating the implementation of control algorithms and simulation systems, on transputer-based architectures. Software tools have been developed, within an environment for automating the partitioning of control algorithms into a number of tasks, and for their allocation on parallel processing transputer-based systems.

A MATLAB-based toolbox has been generated to provide a number of utilities to automate the parallel partitioning of a given control system. The tools facilitate the entry of information arising out of a block diagram description of the system. Systems containing nonlinear elements can also been accommodated. These partitioning tools generate a parallel representation of the original control system as a number of independent tasks. The software performs the partitioning, discretisation and parallel representation of the system as a number state-space equations.

An OCCAM-based toolbox has been developed to automate the mapping of the MATLAB partitioned representation of the system tasks onto a number of transputer-based topologies, using either static or dynamic task allocation strategies. The utilities included in these tools have permitted the evaluation of the allocation strategies, by displaying on-line task allocation, processor activity and execution time.

MATLAB and OCCAM toolboxes developed within EPICAS have been described using the VAP flight control law, chosen as a test algorithm for development purposes and to give insight into the potential performance associated with the strategies of implementation used. This has illustrated that the environment offers a suitable means for automating and evaluating the partitioning and mapping of control systems, for a number of parallel processing topologies. Further, an example of using the toolboxes for simulation purposes has also been illustrated by means of the parallel processing implementation of the models of the flight dynamics of a BAC 1-11 transport aircraft. These models have been partitioned and mapped onto a transputer based system using the MATLAB and OCCAM Toolboxes in order to observe its response to given step inputs.

CHAPTER 5

PERFORMANCE ISSUES: GRANULARITY, TOPOLOGY, MAPPING STRATEGIES

5.1 INTRODUCTION

Parallelism produces a useful benefit when it successfully yields higher performance. When it cannot be used effectively, it adds to the system cost and complexity. This Chapter analyses the performance benefits of the task allocation strategies, described in Chapter 4, running on a number of multiple processor structures. A variety of metrics will be used for measuring the performance of these strategies. A partitioned version of the digital flight control law (VAP Control Law) has been mapped onto a number of parallel processing transputer-based systems, as an example which provides insight into the potential performance improvements and bottlenecks of such systems.

5.2 MEASURING PARALLEL PROCESSOR PERFORMANCE

There are many ways to measure the performance of parallel algorithms running on parallel processing systems. However there is no single suitable metric for absolute performance estimation [74]. It is possible to compare the performance of any two computers when solving the same problem, but the results of this comparison can vary strongly with algorithm and size of problem. Thus a single comparison does not form an adequate basis for evaluating the relative performance of those systems on other applications.

5.2.1 R/C Ratio and Granularity

The performance benefits of using multiple processors depend on the ratio R/C, where R is the length of a run-time quantum and C is the length of communications overhead produced by that quantum [13]. The ratio expresses how much overhead is incurred per unit of computation. When the ratio is very low, it becomes ineffective to use parallelism. When the ratio is very high, parallelism is potentially profitable. Small ratios lead to poor performance because of high communication overheads. Large ratios may usually reflect poor exploitation of parallelism. Then, for maximum performance, it is necessary to balance parallelism against overheads. The ratio R/C is a measure of *task granularity*. In *coarse-grain* parallelism, R/C is relatively high, so each unit of computation produces a relatively small amount of communication. In *fine-grain* parallelism, R/C is low, so there is a relatively large amount of communication and other overheads per unit of computation. Coarse-grain parallelism arises when individual tasks are large and overheads can be amortised over many computational cycles. Fine-grain parallelism takes place when a computational problem is partitioned into increasingly smaller tasks that can run in parallel. Undoubtedly, a problem can be partitioned into the finest possible granularity to create the maximum amount of parallelism. However maximum parallelism generally incurs maximum overhead and this may not lead to the fastest solution. A fine-grain partition that happens to have a low R/C ratio produces poorer performance than a much coarser partition with higher R/C ratio. Hence the much higher parallelism of the fine-grain partition need not produce higher net speed.

5.2.2 Standard Metrics

The most commonly used measurements are the *execution time*, *price/performance*, the *speedup*, and the *efficiency* [75],[76]. The *serial fraction* [77], a new metric that has some advantages over the others, will be used as well. *Execution time* is the effective elapsed time to run a particular job on a given machine. For example, the assignment in the occam fragment below:

```
REAL32 a, b, c:
SEQ
  ...
  a := b * c
  ...
```

is executed by a transputer T414-20 in about 11 µs provided all the code and variables are in internal RAM. On the T800-20 the process above would take only 1.45 µs. By comparison, the same assignment on an 8 Mhz Intel 80286/80287 combination would take about 31 µs. *Price/performance* of a parallel system is the elapsed time for a program divided by the cost of the machine that ran the job. This is important if there are a number of machines that are "fast enough". Given a fixed amount of money, it may be advantageous to buy a number of "slow" machines rather than one fast machine. These two measurements are used to help to decide which machine to buy. Once the machine is available, *speedup* and *efficiency* are often used to measure how effectively the machine is being used.

Speedup is generally measured by running the same program on a varying number of processors. Speedup is then the elapsed time needed by *1* processor divided by the time needed on *p* processors:

$$s = \frac{T[1]}{T[p]}. \qquad (5.1)$$

In studying algorithms for parallel processors, if a system A gives a higher speedup than a system B for the same program, then A provides better support for parallelising this program than does system B. An example of such support is the presence of more processors. As the speedup of a system is highly dependent on the number of processors, then this parameter must be also included. By scaling speedup by the number of processors used, a more representative performance measure is achieved. This metric is called *efficiency* and is defined as

$$e = \frac{T[1]}{pT[p]} = \frac{s}{p}. \tag{5.2}$$

Efficiency close to unity suggests that the hardware is being used effectively; low values of efficiency means that resources are being wasted.

Each of these metrics has its disadvantages. In fact, there is important information that cannot be obtained even by looking at all of them. Adding processors should reduce the execution time, but by how much? *Speedup* provides this data. Speedup close to linear is good, but how close to linear is good enough? *Efficiency* can reveal how close the user is getting to the best the hardware can perform, but when the efficiency is not particularly high, what is the cause? The metric called *serial fraction*, is intended to answer these questions.

The *serial fraction* is derived from Amdahl's Law which in its simplest form says that

$$T[p] = T_s + \frac{T_p}{p}, \tag{5.3}$$

where T_s is the time taken by the part of the program that must be run serially and T_p is the time in the parallelisable part. Considering $T[1] = T_s + T_p$. If we define the serial fraction, $f = T_s / T[1]$ then Eq. (5.3) can be written as

$$T[p] = T[1]f + \frac{T[1](1-f)}{p}, \tag{5.4}$$

From Eqs. (5.1) and (5.4), the serial can be expressed in terms of s and p, namely

$$f = \frac{1/s - 1/p}{1 - 1/p}, \quad \text{for } p > 1 \tag{5.5}$$

The value of f is useful because Eq. (5.3) is incomplete. First, it assumes that all processors compute for the same amount of time and the work is perfectly load balanced. If the load is not perfectly balanced, the speedup will be reduced giving a larger serial fraction. Secondly, is ignored in Eq. (5.3), the overhead of synchronising processors. Load-balancing effects are likely to result in an irregular change in f as p increases. For example, if there are 12 pieces of work to do that take the same amount of time, there will be perfect load balancing for 2, 3, 4, 6, and 12 processors, but less than perfect load balancing for other values of p. Since a larger load imbalance results in a larger increase of f it is possible to identify problems not apparent from speedup or efficiency. On the other hand, the overhead of synchronising processors is a monotonically increasing function of p. Since increasing overhead decreases the speedup, this effect results in a smooth increase in the serial fraction f as p increases. Smoothly increasing f is a warning that the granularity of the parallel tasks is too fine.

In summary, the serial fraction can provide information not revealed by other commonly used metrics. Execution time, speedup and efficiency vary as the number of processors increases. In an ideal system, with no serial component and even load balancing with no communications overhead, the serial fraction would be zero. In a practical system, therefore, we want the serial fraction to be close to zero and, also to remain nearly constant as the number of processors increases. Deviations from a constant level will provide indication of a deteriorating performance as the number of processors increases.

5.2.3 Scaled Metrics

The metrics presented so far refer to problems of fixed size. Current approaches [78], claim that in practice, *the problem size scales with the number of processors*. Then, performance issues should be measured by scaling the problem to the number of processors, not by fixing problem size. The user will increase the problem size to keep the elapsed run-time more or less constant. As the problem size grows, we should find that the fraction of the time spent executing serial code decreases, leading us to predict a decrease in the measured serial fraction. If we assume that the serial time and overhead are independent of problem size:

$$T[p,k] = T_s + \frac{kT_p}{p}, \qquad (5.7)$$

where $T[p,k]$ is the time to run the program on p processors for a problem needing k times more processing. Here k is the scaling factor ($k=1$ when $p=1$). The speedup, s_k, for the scaled problem must account for additional processing that must be done to solve a larger problem:

$$s_k = \frac{kT[1,1]}{T[p,k]}. \qquad (5.8)$$

The scaled efficiency is then $e_k = s_k/p$ \hfill (5.9)

and the scaled serial fraction becomes

$$f_k = \frac{1/s_k - 1/p}{1 - 1/p}. \qquad (5.10)$$

According with the previous definitions we see that $f=kf_k$ which under ideal circumstances would remain constant as p increases.

5.3 PROCESSOR FARM PERFORMANCE

The standard metrics, described in Sections 5.2.2, have been used for measuring the performance of the transputer-based processor farm model described in Chapter 4.

5.3.1 Farm Using a Linear Topology

The *execution time* of the linear arrangement of the farm was measured using the real-time clock facility of the transputers. Other metrics (*speedup*, *efficiency* and *serial fraction*) have been calculated, starting with the execution time measurements. Execution time has been determined by calculating the difference of *timer* input statements placed within the master process. Measurements were carried out varying the number of workers in the farm, from 1 up to 6 workers, in the test case presented. Figure 5.1 shows the resulting time performance of this array for two types of transputers (T414-15 and T800-20). The *total time*, in each case, is the interval in which the master executes a complete iteration (allocates tasks, receives results and computes the global system outputs). The time during which the processor farm is actively computing results, called *calculation time*, is determined by the difference between the *total time* and the period when the system is *communicating* data or is *idle*. This latter period was obtained running a modified version of **PROGRAM** *farmline* where the *calculate* process on the worker process has been commented out. The *speedup*, *efficiency*, and the *serial fraction*, for this farm model have been calculated according to the Eqs. (5.1), (5.2) and (5.5) respectively. The results are summarised in Table 5.1. Looking at Table 5.1, the linear farm of transputers T414-15 (T4), shows a modest execution time speedup ranging from 1.273 to 1.422, as additional processing power is introduced. However, no extra improvement can be obtained for more than 5 workers. Furthermore, for 6 workers or more the execution time is longer than the previous cases. The efficiency values, which range from 0.64 to 0.22, indicate that a low utilisation of resources is being made. This aspect also can be observed in Figure 5.1, comparing the calculation time against the total execution time. Since increasing overhead decreases the speedup, this effect results in a smooth increase in the serial fraction f as p increases. This gradual increase of f is an indication that the granularity of the parallel tasks is too fine. In the current realisation, the scheduling action performed by the master is only a list scheduling of independent tasks, and thus does not present a significant overhead to the performance. Communications overhead is the important component and is exacerbated when the linear array is long. This contrasts with the findings of related research which

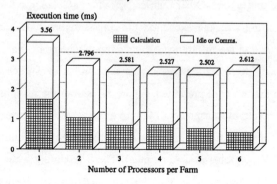

(a)

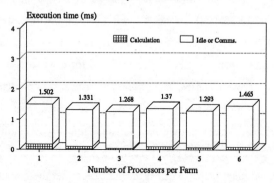

(b)

Figure 5.1 Performance of a processor farm -linear array.
(a) T414-15 transputer implementation
(b) T800-20 transputer implementation.

System	p	Time (ms)	s	e	f
FarmLine-T4	1	3.56	-	-	-
FarmLine-T4	2	2.796	1.273	0.64	0.571
FarmLine-T4	3	2.581	1.379	0.46	0.588
FarmLine-T4	4	2.527	1.408	0.35	0.614
FarmLine-T4	5	2.502	1.422	0.28	0.629
FarmLine-T4	6	2.612	1.362	0.22	0.681
FarmLine-T8	1	1.502	-	-	-
FarmLine-T8	2	1.331	1.128	0.56	0.773
FarmLine-T8	3	1.268	1.185	0.39	0.766
FarmLine-T8	4	1.376	1.096	0.27	0.883
FarmLine-T8	5	1.293	1.162	0.23	0.826
FarmLine-T8	6	1.465	1.025	0.17	0.971

Note: p=#processors, s=speedup, e=efficiency, f=serial fraction.

Table 5.1 Summary of Processor Farm Performance
Case Study: VAP Control Law

applied process farming to the calculation of the inverse dynamics for a robot manipulator [32], where the scheduling operation took almost 70 % of the total execution time.
Also, it should be noted that the scale of computation in both cases is quite different (about 100 flops, in the flight control example, against 1500 flops in the inverse dynamics case).

Using T800 transputers, obviously shows a considerable improvement in execution times when compared with the T414 implementation. The typical T800 floating point operations, used in this application example for the calculation processes, are about 500 ns [70], whereas the operations in the T414 transputer have an average value of 10µs). Here the execution time ranges from 1.502 ms to 1.268 ms. However, the maximum speedup achieved was $s = 1.185$, using 3 workers. The R/C ratio has decreased in this case as R became smaller. Because this ratio expresses a higher overhead per unit of computation, it thus becomes ineffective to use parallelism. The lower processor activity is reflected in the values of efficiency. The high and smoothly changing figures in f also reveal the communication overheads as the number of processors grows.

The key to the success of the processor farm model is the requirement that the time consumed by the processors in performing communication is small relative to that expended for computation. However, when this requirement is not matched and a significant overhead on communicating data items exists, this linear array may be unsuitable for any more than a small number of processors in the farm due to the limited throughput of this topology. In order to increase the efficiency of the farm approach, further improvements have been made to the original approach.

The first of these improvements, described in Section 5.3.2, includes the use of task buffering within the workers, so that a new task can be passed to the application (calculating process), awaiting execution, with a minimal latency when it becomes idle. A second refinement is described in Section 5.3.3. This comprises an extended model of the processor farm using three strings of workers arranged around a central master, thus reducing the number of channels through which messages must travel and increasing the bandwidth of data through the master.

5.3.2 Farm Using Buffers

In the basic processor farm, a worker cannot continuously execute tasks. When a task has been processed, it is passed to the *feedback* process which returns it to the master. This, in turn, sends a new task to the network to replace the one which has been completed. However, rather than allowing a worker to idle whilst the farmer issues new work, the worker process may buffer an extra task so that it may proceed immediately with this new task once it has completed the previous work. This scheme strives to keep the workers constantly busy. Thus it was convenient to include a task buffer, between the *throughput* and the *calculate* processes, within each worker. Each worker processor may therefore be allocated a second task which is ready to proceed as soon as the current task is complete. The *worker* process in this buffered version of the farm called **PROGRAM** *farmbuff* now includes a *buffer* process, see Figure 5.2.

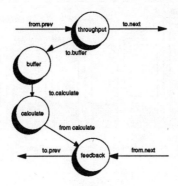

Figure 5.2 The component processes of a buffered worker.

The *master* has the same structure as in the original version, but the *coordinator* now initialises the farm by issuing *2 * number.of.workers* data packets to the work force. This data just fills up each worker and each buffer. Figure 5.3 shows the resulting execution time performance of this array for transputers T414-15 and T800-20. The *speedup, efficiency,* and the *serial fraction* figures are summarised in Table 5.2. Looking at Table 5.2, using 3 transputers T414-15 (T4), task buffering achieves a maximum execution time speedup *s* of 1.41, similar to that obtained using the original unbuffered farm with 5 transputers (1.42), see Table 5.1. However, for a larger number of workers the performance of the buffered implementation becomes inferior to the former version, because even if a worker becomes idle and is ready to perform a task, it cannot execute a task that has been buffered to another worker. This overhead leads to a reduction in the speedup, giving a larger value of serial fraction *f*, when more than 3 transputers were used. This problem is exacerbated in the T800 version of the buffered approach. In this particular case study, it was not possible to achieve any practical speedup; on the contrary, the execution time was increased as additional

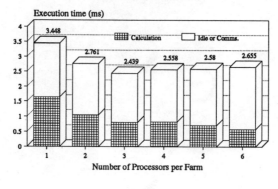

(a)

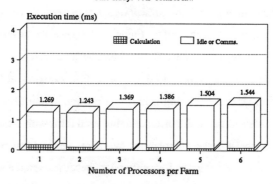

(b)

Figure 5.3 Performance of a processor farm using task buffering.

    (a) T414-15 transputer implementation.

    (b) T800-20 transputer implementation.

System	p	Time (ms)	s	e	f
Farmbuff-T4	1	3.448	-	-	-
Farmbuff-T4	2	2.761	1.249	0.62	0.601
Farmbuff-T4	3	2.439	1.414	0.47	0.561
Farmbuff-T4	4	2.558	1.348	0.38	0.656
Farmbuff-T4	5	2.580	1.336	0.27	0.686
Farmbuff-T4	6	2.655	1.299	0.22	0.724
Farmbuff-T8	1	1.269	-	-	-
Farmbuff-T8	2	1.243	1.021	0.51	0.959
Farmbuff-T8	3	1.369	0.927	0.31	1.118
Farmbuff-T8	4	1.386	0.916	0.23	1.122
Farmbuff-T8	5	1.504	0.844	0.17	1.231
Farmbuff-T8	6	1.544	0.822	0.14	1.260

Note: p=#processors, s=speedup, e=efficiency, f=serial fraction.

Table 5.2 Performance Summary of Processor Farm Using Task Buffering Case Study: VAP Control Law.

T800 transputers were introduced. Thus, the task buffering version of the processor farm, realised with T414 transputers provided a more efficient implementation than the unbuffered approach, but only for a small number of workers, 3 in this case.

5.3.3 Farm Using an Extended Topology

A linear arrangement may be unsuitable for any more than a small number of processors in the farm due to its limited throughput. A message may require a number of $O(N)$ propagations to reach its destination in a network of N processors and each must route $O(N)$ messages). Further research has been carried out to investigate schemes for scheduling tasks to the processors in the farm which keep the overhead of communicating data to a minimum [27], [79]. One of these schemes is to increase the effective throughput by adopting different topologies of processors (e.g. tree, star, etc.). Based on these strategies, the original linear farm model has been extended by configuring the system as three chains of worker processors arranged around a central master, see Figure 5.4.

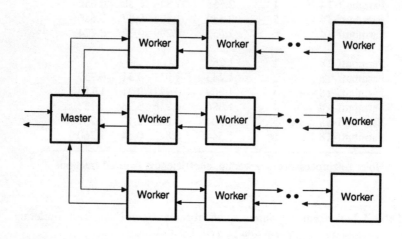

Figure 5.4 Processor Farm -Extended Topology

Three chains are used due to the fact that transputers have only four links and the fourth link of the master is interfaced to the monitor process in the transputer development system for input/output. This approach required a more sophisticated task allocation strategy. However, a reduction in communication overhead was expected, because in this extended approach there is a shorter path between the controller and any worker. Within this augmented model, the *worker* process maintained the same software structure as in the original linear and non-buffered array, but the *master*, was modified in order to schedule tasks in parallel to three strings of workers. The summarised component processes of the extended *master* process are

-- **sender**

... receive inputs & initial tasks, send these to branches

WHILE tasks.are.available
    ALT
        ... free worker from branch 1?, send new task to branch 1
        ... free worker from branch 2?, send new task to branch 2
        ... free worker from branch 3?, send new task to branch 3

-- coordinator
... issue initial tasks to workers via sender
WHILE tasks.are.available
    ALT
        ... free worker from branch 1?, route identifier to sender
        ... free worker from branch 2?, route identifier to sender
        ... free worker from branch 3?, route identifier to sender

-- receiver
WHILE tasks.are.available
    ALT
        ... worker branch 1?, identifiers to coord. & results to moni.
        ... worker branch 1?, identifiers to coord. & results to moni.
        ... worker branch 1?, identifiers to coord. & results to moni.

The whole structure of the extended processor farm can be expressed as follows

PAR
    master (forward1[0], return1[0], forward2[0], return2[0], forward3[0], return3[0])
    PAR i = 1 **FOR** workers.per.branch1
        worker (forward1[i-1], return1[i-1], forward1[i], return1[i])
    PAR i = 1 **FOR** workers.per.branch2
        worker (forward2[i-1], return2[i-1], forward2[i], return2[i])

PAR i = 1 **FOR** workers.per.branch3
    worker (forward3[i-1], return3[i-1], forward3[i], return3[i])

This extended model of the processor farm was implemented in an occam **PROGRAM** called *farmtrip* and used to execute the VAP control law example. Figure 5.5 shows the architectural array and the task activity of each worker for the extended farm (using 3 and 6 workers).

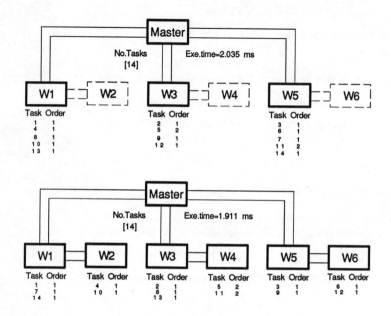

Figure 5.5 Processor activity -Extended farm topology.

The execution time performance of the extended array is shown in Figure 5.6. The values of the corresponding metrics are summarised in Table 5.3. From Figure 5.6, as predicted, the extended topology consistently performed better than the linear farm for the same number of workers. The extended farm topology, for this particular case study, using T414 transputers achieved a lower execution time, ranging from 2.035 to 1.911 ms (for 3 and 6 workers

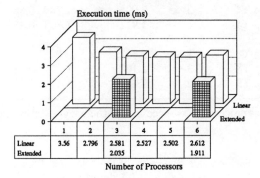

(a)

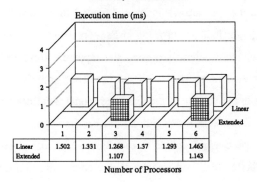

(b)

Figure 5.6 Performance of the extended farm topology.

    (a) T414-15 transputer implementation.

    (b) T800-20 transputer implementation.

System	p	Time (ms)	s	e	f
Farmtrip-T4	1	3.56	-	-	-
Farmtrip-T4	3	2.035	1.749	0.58	0.357
Farmtrip-T4	6	1.911	1.863	0.31	0.444
Farmtrip-T8	1	1.502	-	-	-
Farmtrip-T8	3	1.107	1.357	0.45	0.606
Farmtrip-T8	6	1.143	1.314	0.22	0.713

Note: p=#processors, s=speedup, e=efficiency, f=serial fraction.

Table 5.3 Performance Summary of the Extended Processor Farm Case Study: VAP Control Law.

respectively), than the equivalent number of workers in the linear topology (2.581 to 2.612 ms). This represents a not inconsiderable reduction, considering that the same number of workers are used in both cases. At the same time, from Table 5.3 it can be seen that only a modest improvement in the speedup was achieved from 3 to 6 workers. Since a larger load imbalance results in a larger increase in f, the large increase in the value of f for 6 workers suggests that a load balance problem is present (not apparent from speedup or efficiency). This problem becomes evident by observing the processor activity showed in Figure 5.5. Using T800 transputers in the version of the extended model, a further reduction in execution time is attained using 3 workers only. But no further improvement is achieved using 6 workers (when an extra processor is added to each line) and a loss of efficiency is apparent. The loss of efficiency is due to a load imbalance and communication overheads.

5.4 PROCESSOR STAR PERFORMANCE

The performance of the star topology arrangement has been measured using the same standard metrics for the process farm case. The *execution time* of the star array was evaluated using the real-time clock facility of the transputers. The other metrics (*speedup*, *efficiency* and *serial fraction*) have been calculated using the execution time measurements.

The execution time has been determined by calculating the differences of *timer* input statements placed within the central process. Measurements were carried out, varying the number of workers in the array, from 0 (when the master performs the whole algorithm) up to 3 workers, in the test case presented. Figure 5.7 shows the execution time performance of this array, using T414-15 and T800-20 transputers. Here, *total time*, is defined as the interval in which the master executes a complete iteration (performing its own tasks, receiving task results from the workers and performing the supplementary calculation to generate the global system outputs). The time that the array is actively computing results, *calculation time*, is determined by the difference between the *total time* and the interval when the system is *communicating* data or *idle*. This final period was obtained running a version of **PROGRAM procstar**, for which all the calculations had been commented out, on the master and worker processes. The *speedup*, *efficiency*, and the *serial fraction* figures are summarised in Table 5.4. Comparing the values shown in Table 5.4 with the values of Tables 5.1 - 5.3 of previous implementations, it is clear that the star topology exhibited a good performance. There is a significant increase in the execution time speedup as additional processing power is introduced but at the expense of a reduction in efficiency. The increasing value of the serial fraction f, suggests that the loss of efficiency is due to overhead in synchronising processors caused by a load imbalance of tasks, coupled with communication overheads. Nevertheless this approach performs better than other implementations (dynamic allocation) for the same number of processors.

The use of T800 transputers, similarly revealed a considerable improvement in the execution times, when compared with the T414 implementation, but smaller values of efficiency can be noted. (The R/C ratio of the tasks is diminished because the run time R of the tasks diminished due to the floating point unit on the T800). This is also manifested in the values of f, which are larger than those obtained by the T414 transputer implementation.

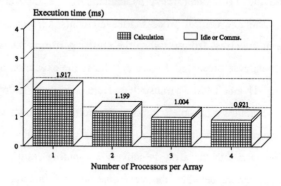

(a)

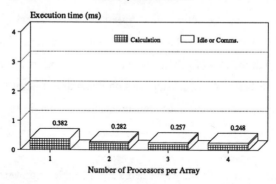

(b)

Figure 5.7 Performance of processor star array.

(a) T414-15 transputer implementation.

(b) T800-20 transputer implementation.

System	p	Time (ms)	s	e	f
Procstar-T4	1	1.917	-	-	-
Procstar-T4	2	1.199	1.598	0.8	0.252
Procstar-T4	3	1.004	1.909	0.64	0.286
Procstar-T4	4	0.921	2.081	0.52	0.307
Procstar-T8	1	0.382	-	-	-
Procstar-T8	2	0.282	1.354	0.68	0.477
Procstar-T8	3	0.257	1.486	0.5	0.509
Procstar-T8	4	0.248	1.54	0.39	0.532

Table 5.4 Performance Summary of Processor Star Topology Case Study: VAP Control Law.

5.5 GRANULARITY ISSUES

The performance analysis described in Sections 5.3 - 5.5 has explored many facets of the task allocation strategies. However, all the analysis so far refers to a particular problem of fixed size (the parallel implementation of the VAP control law). The results of this study, however, can vary strongly with the algorithm and size of the problem. This consideration leads us to apply the software tools to more complex problems in order to analyse and evaluate the performance of the tools in a more general way.

In order to generate problems of a greater complexity, where the allocation tools (both static and dynamic approaches) could be applied and evaluated, two different strategies were implemented. The first one, considered the effects on performance when the algorithm size is increased and a larger number of tasks was executed by a system, keeping the original scale of granularity. The second approach, deals with the effect of increasing the order of complexity of the tasks, by means of associating small grains together on single processors (augmenting the grain size). This was done in order to provide better estimates of speed improvements for future, much larger problems.

5.5.1 Performance of Models with a Varying Number of Tasks

This Section investigates the effect on the performance of the allocation strategies when the number of tasks to be executed is increased. This effect has been considered for both the static and dynamic task allocation cases. The original database corresponding to the VAP control algorithm, was duplicated, and then triplicated. Thus, two new simulated algorithms were obtained: the first one with 28 tasks, and the second, with 42 tasks. The processor farm model, implemented in the occam **PROGRAM** called *farmline*, was used to allocate dynamically the task sets of each algorithm. Figure 5.8, depicts the execution time performance of this approach using T414-15 and T800-20 transputers. Figures 5.9 - 5.11 present the associated values of *speedup*, *efficiency*, and *serial fraction* for the three cases (14, 28 and 42 tasks).

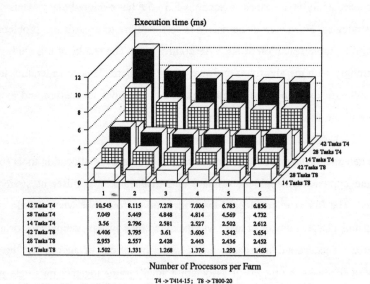

Figure 5.8 Execution time for processor farm topology: 14, 28 and 42 tasks.

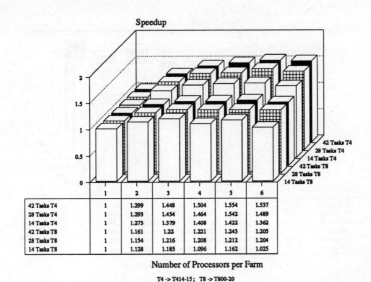

	1	2	3	4	5	6
42 Tasks T4	1	1.299	1.448	1.504	1.554	1.537
28 Tasks T4	1	1.293	1.454	1.464	1.542	1.489
14 Tasks T4	1	1.273	1.379	1.408	1.422	1.362
42 Tasks T8	1	1.161	1.22	1.221	1.243	1.205
28 Tasks T8	1	1.154	1.216	1.208	1.212	1.204
14 Tasks T8	1	1.128	1.185	1.096	1.162	1.025

Number of Processors per Farm

T4 -> T414-15; T8 -> T800-20

Figure 5.9 Speedup for processor farm topology: 14, 28 and 42 tasks.

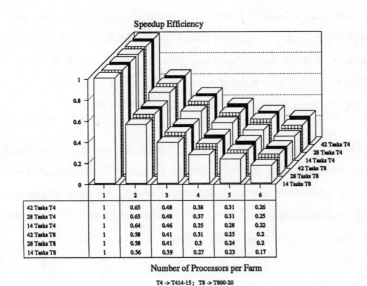

	1	2	3	4	5	6
42 Tasks T4	1	0.65	0.48	0.38	0.31	0.26
28 Tasks T4	1	0.65	0.48	0.37	0.31	0.25
14 Tasks T4	1	0.64	0.46	0.35	0.28	0.22
42 Tasks T8	1	0.58	0.41	0.31	0.25	0.2
28 Tasks T8	1	0.58	0.41	0.3	0.24	0.2
14 Tasks T8	1	0.56	0.39	0.27	0.23	0.17

Number of Processors per Farm

T4 -> T414-15; T8 -> T800-20

Figure 5.10 Efficiency for processor farm topology: 14, 28 and 42 tasks.

Performance issues: granularity, topology, mapping strategies   *Chapter 5*

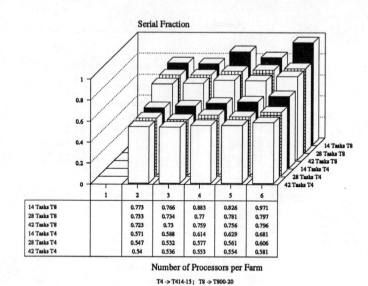

Figure 5.11 Serial fraction values for processor farm topology: 14, 28 and 42 tasks.

The values displayed in Figure 5.8, show an obvious increase in the execution time. This increment is proportional to the algorithm size or complexity, because, keeping the granularity at the same scale, a larger number of tasks (plus the associated routing process) must be executed. Thus a similar performance is anticipated for all these cases. The values in Figures 5.9 - 5.11 confirm this expectation, for the 28 and 42 tasks cases.

Looking at the T414 implementation values in Figures 5.9 and 5.10, it is possible to note smaller values of speedup and efficiency, for the case of 14 tasks, compared with those for 28 and 42 tasks, where very similar values were obtained. The irregular change and higher values, in the corresponding serial fraction data for the case of 14 tasks (lower half of values presented in Figure 5.11), were the result of load-balancing effects.

Performance issues: granularity, topology, mapping strategies   *Chapter 5*

A very similar situation was noticed for the implementation using transputers T800, the main difference being the smaller execution time scale achieved by this implementation processor, but at the cost of an inferior efficiency.

For the processor star approach, the occam **PROGRAM** called *procstar*, was used to allocate statically the algorithms. The execution time performance of this strategy is shown in Figure 5.12. The associated metrics are displayed in Figures 5.13 - 5.15. As in the previous implementation, there was an increase in the algorithm execution time, proportional to its size or complexity. However for this implementation the performance of the three algorithms was similar, this could be noted by inspecting the values of speedup and efficiency. The regular serial fraction values also indicated that there was a good load balance in the cases analysed.

We conclude, therefore, that the larger the number of tasks to be executed, the greater the potential for load balancing and, thus, efficient use of the parallel processing system.

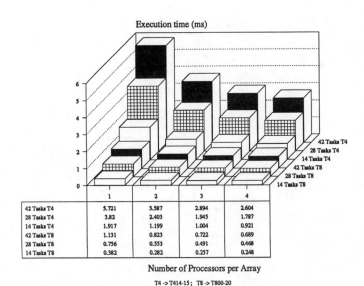

Figure 5.12 Execution time for processor star topology: 14, 28 and 42 tasks.

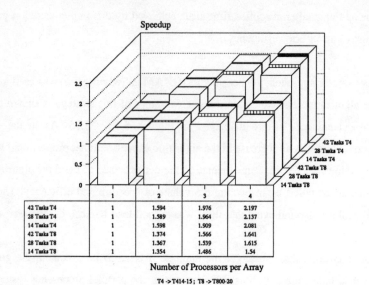

Figure 5.13 Speedup for processor star topology: 14, 28 and 42 tasks.

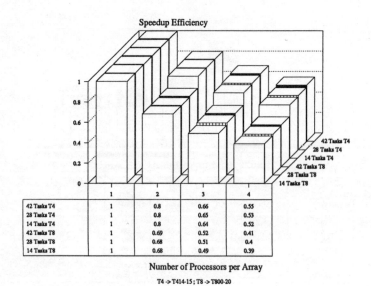

Figure 5.14 Efficiency for processor star topology: 14, 28 and 42 tasks.

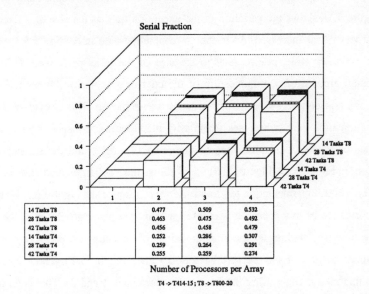

Figure 5.15 Serial fraction values of processor star topology: 14, 28 and 42 tasks.

5.5.2 Performance of Models with a Higher Task Granularity

This section investigates the effect, on the performance of the allocation strategies when the complexity of the tasks is augmented, by means of increasing the granularity, while keeping the number of tasks constant. In order to augment the grain size of the original tasks, corresponding to the VAP control algorithm, additional processing was introduced into the task model executed by each processor on the array, simply by replication of code. The resulting size of the task models ranged from *size 2* to *size 8*. A size 2 task model, for example, will execute twice the arithmetic than the reference model (*size 1*). A *size 3* task will execute three times the calculation, and so on. The processor farm **PROGRAM** *farmline* was used to evaluate its performance on the computation of this augmented algorithms. In order to achieve this purpose, the task model, mapped onto the *worker* process, was modified to scale the task to the different sizes, as mentioned above. Figure 5.16, depicts the execution time performance of this approach using T414-15 transputers. Figures 5.17-5.19 present the values of *speedup*, *efficiency*, and *serial fraction* for the eight different sizes of the task models. Figure 5.16 shows the variation of the execution time as the size of a fixed number of tasks (14 tasks) was increased. As in the corresponding case in Section 5.5.1, there is an increase in execution time, proportional to the size of the tasks performed. However, the relatively small amount of communication of the coarser granularity (higher R/C) reduced overheads, and consequently a better performance was achieved. This is evident by looking at the execution time data (when task granularity increases), and inspecting the associated higher values of speedup and efficiency in Figures 5.17 - 5.18, and the smaller values of serial fraction, in Figure 5.19. Figures 5.20 - 5.23 shows the T800 implementation of the same problem. The T800 implementation led to a faster execution of the algorithms. However the overall performance of this approach was inferior than that using processor of the type T4. This can be noted by looking at the smaller values of speedup and efficiency and higher values of serial fraction. Note the serial fraction values in Fig. 5.23. There is an uneven variation in this metric when using the finer grain tasks (size 1 and 2). This is an indication of the granularity mismatch experienced between these task sizes and the T8 implementation.

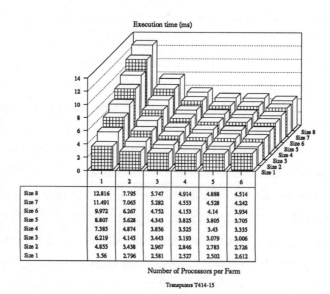

Figure 5.16 Execution time for a processor farm using T4's: 14 tasks, grain size (1 - 8).

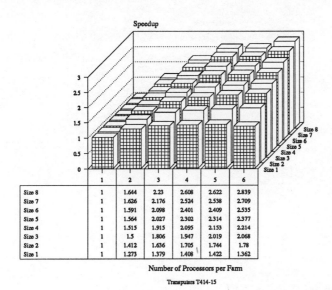

Figure 5.17 Speedup for processor farm using T4's: 14 tasks, grain size (1 - 8).

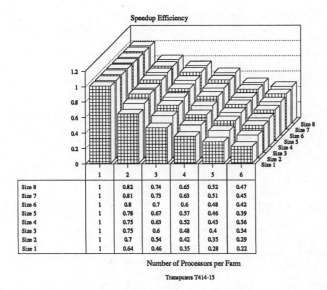

Figure 5.18 Efficiency for processor farm using T4's: 14 tasks, grain size (1 - 8)

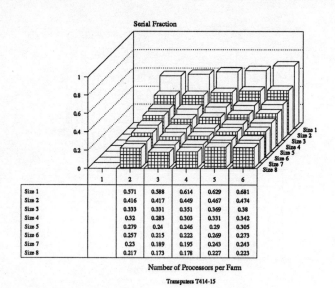

	1	2	3	4	5	6
Size 1		0.571	0.588	0.614	0.629	0.681
Size 2		0.416	0.417	0.449	0.467	0.474
Size 3		0.333	0.331	0.351	0.369	0.38
Size 4		0.32	0.283	0.303	0.331	0.342
Size 5		0.279	0.24	0.246	0.29	0.305
Size 6		0.257	0.215	0.222	0.269	0.273
Size 7		0.23	0.189	0.195	0.243	0.243
Size 8		0.217	0.173	0.178	0.227	0.223

Figure 5.19 Serial fraction values for processor farm using T4's: 14 tasks, grain size (1 - 8).

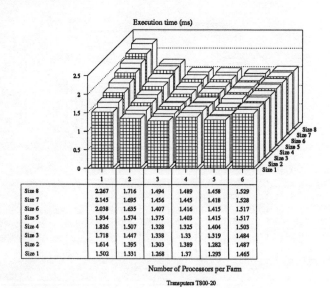

	1	2	3	4	5	6
Size 8	2.267	1.716	1.494	1.489	1.458	1.529
Size 7	2.145	1.695	1.456	1.445	1.418	1.528
Size 6	2.038	1.635	1.407	1.416	1.415	1.517
Size 5	1.934	1.574	1.375	1.403	1.415	1.517
Size 4	1.826	1.507	1.328	1.325	1.404	1.503
Size 3	1.718	1.447	1.338	1.33	1.319	1.484
Size 2	1.614	1.395	1.303	1.389	1.282	1.487
Size 1	1.502	1.331	1.268	1.37	1.293	1.465

Figure 5.20 Execution time for a processor farm using T8's: 14 tasks, grain size (1 - 8).

Performance issues: granularity, topology, mapping strategies   Chapter 5

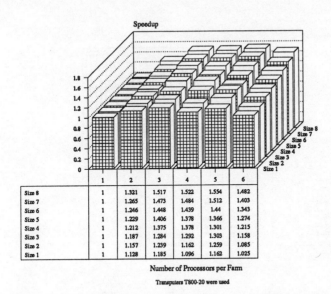

	1	2	3	4	5	6
Size 8	1	1.321	1.517	1.522	1.554	1.482
Size 7	1	1.265	1.473	1.484	1.512	1.403
Size 6	1	1.246	1.448	1.439	1.44	1.343
Size 5	1	1.229	1.406	1.378	1.366	1.274
Size 4	1	1.212	1.375	1.378	1.301	1.215
Size 3	1	1.187	1.284	1.292	1.303	1.158
Size 2	1	1.157	1.239	1.162	1.259	1.085
Size 1	1	1.128	1.185	1.096	1.162	1.025

Number of Processors per Farm

Transputers T800-20 were used

Figure 5.21   Speedup for processor farm using T8's: 14 tasks, grain size (1 - 8).

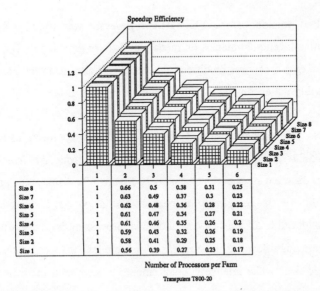

	1	2	3	4	5	6
Size 8	1	0.66	0.5	0.38	0.31	0.25
Size 7	1	0.63	0.49	0.37	0.3	0.23
Size 6	1	0.62	0.48	0.36	0.28	0.22
Size 5	1	0.61	0.47	0.34	0.27	0.21
Size 4	1	0.61	0.46	0.35	0.26	0.2
Size 3	1	0.59	0.43	0.32	0.26	0.19
Size 2	1	0.58	0.41	0.29	0.25	0.18
Size 1	1	0.56	0.39	0.27	0.23	0.17

Number of Processors per Farm

Transputers T800-20

Figure 5.22   Efficiency for processor farm using T8's: 14 tasks, grain size (1 - 8)

Performance issues: granularity, topology, mapping strategies *Chapter 5*

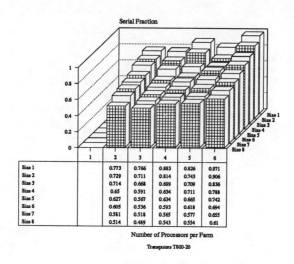

Figure 5.23 Serial fraction values for processor farm using T8's: 14 tasks, grain size (1 - 8).

Similarly, a modified version of the **PROGRAM** *procstar* was used to evaluate the performance of static allocation strategies on models of coarser granularity. The task model allocated on the *worker* process was modified to scale its task size, as in the processor farm approach, by increasing the arithmetic that must be done by each task.

Figure 5.24, depicts the execution time performance of this approach using T414-15 transputers. Figures 5.25 - 5.27 present the associated values of *speedup*, *efficiency*, and *serial fraction* for the eight different sizes of the task models. Figure 5.24 displays the profile of the execution time, as the size or complexity of the tasks (14 tasks) was increased. Execution time speedup was achieved in all the cases, as additional processing power was added to the system. This was more significant as the size of the task models, mapped on the array, was increased, see Figure 5.25. The relatively small amount of communication of the coarser granularity (higher R/C) reduced overheads, and consequently improved the performance.

Performance issues: granularity, topology, mapping strategies   *Chapter 5*

At the same time, according to Figure 5.26, a higher efficiency, proportional to the task size, could be observed. The very small values of the serial fraction (Figure 5.27) for the same situation, also indicate that a more efficient implementation was achieved as the size of the grain was increased.

The T800 implementation for the same problem is displayed in Figures 5.28 - 5.31. This implementation again led to a significant reduction in execution time for the models presented. However the overall performance of this implementation, as in previous cases already considered, is inferior to that using T414 processors. The corresponding smaller figures of speedup and efficiency and relatively higher values of the serial fraction, for this implementation, support this observation.

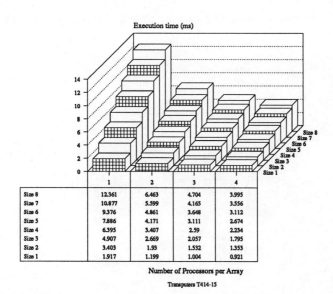

Figure 5.24 Execution time for a processor star using T4's: 14 tasks, grain size (1 - 8).

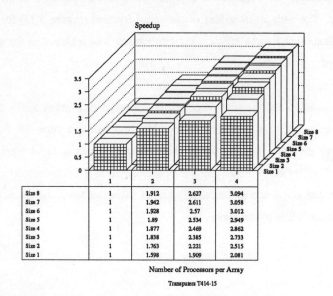

Figure 5.25 Speedup for processor star using T4's: 14 tasks, grain size (1 - 8).

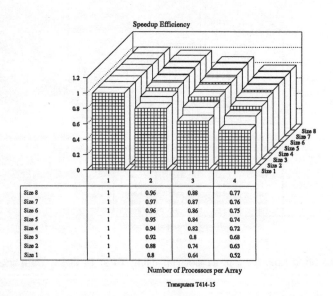

Figure 5.26 Efficiency for processor star using T4's: 14 tasks, grain size (1 - 8)

Performance issues: granularity, topology, mapping strategies   *Chapter 5*

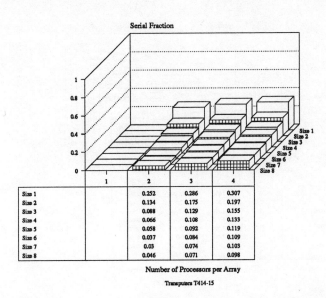

	1	2	3	4
Size 1		0.252	0.286	0.307
Size 2		0.134	0.175	0.197
Size 3		0.088	0.129	0.155
Size 4		0.066	0.108	0.133
Size 5		0.058	0.092	0.119
Size 6		0.037	0.084	0.109
Size 7		0.03	0.074	0.103
Size 8		0.046	0.071	0.098

Number of Processors per Array

Transputers T414-15

Figure 5.27 Serial fraction values of processor star using T4's: 14 tasks, grain size (1 - 8).

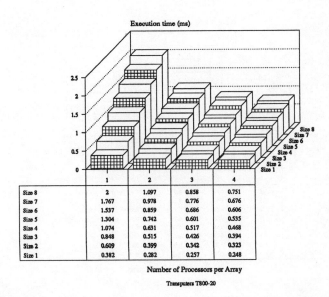

	1	2	3	4
Size 8	2	1.097	0.858	0.751
Size 7	1.767	0.978	0.776	0.676
Size 6	1.537	0.859	0.686	0.606
Size 5	1.304	0.742	0.601	0.535
Size 4	1.074	0.631	0.517	0.468
Size 3	0.848	0.515	0.426	0.394
Size 2	0.609	0.399	0.342	0.323
Size 1	0.382	0.282	0.257	0.248

Number of Processors per Array

Transputers T800-20

Figure 5.28 Execution time for a processor star using T8's: 14 tasks, grain size (1 - 8).

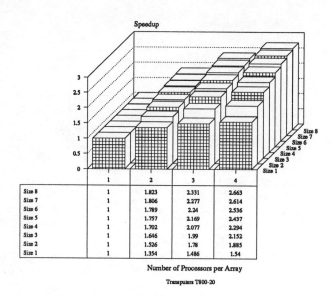

Figure 5.29 Speedup for processor star using T8's: 14 tasks, grain size (1 - 8).

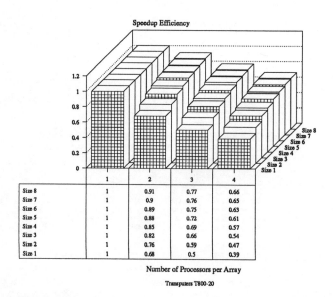

Figure 5.30 Efficiency for processor star using T8's: 14 tasks, grain size (1 - 8)

Performance issues: granularity, topology, mapping strategies   *Chapter 5*

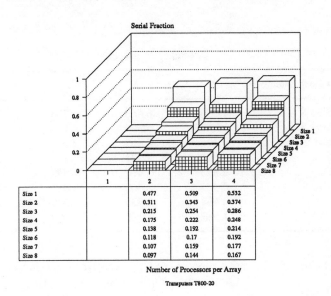

Figure 5.31 Serial fraction values for processor star using T8's: 14 tasks, grain size (1 - 8).

5.5.3 Tools Performance on Scaled-sized Models

Modern approaches for measuring parallel processor performance (Section 5.2.3), states that performance should be measured by scaling the problem to the number of processors, not by fixing the problem size. Given a more powerful processing capacity, the problem generally should expand to make use of the increased facilities, keeping the execution time constant. The definition of speedup now considers the additional processing that must be done to solve a larger problem. The relevance of this to digital control is that parallel processing may be used to perform additional useful processing within the sample window. This extra processing may handle control algorithms of increased complexity, new functional components, incorporation of a system model for condition monitoring purposes or software to handle possible fault scenarios. In order to provide an example of this new approach for measuring performance, the execution time values of Figures 5.24 and 5.28 (formerly evaluated in

Section 5.5.2) were considered. The execution time data is summarised in Tables 5.5 and 5.6, respectively.

T414-15	Number of processors per array			
Tasks Size	1	2	3	4
Size 8	12.361	6.463	4.704	3.995
Size 7	10.877	5.599	4.165	3.556
Size 6	9.376	4.861	3.648	3.112
Size 5	7.886	4.171	3.111	2.674
Size 4	6.395	3.407	2.590	**2.234**
Size 3	4.907	2.669	**2.057**	1.795
Size 2	3.403	**1.930**	1.532	1.353
Size 1	**1.917**	1.199	1.004	0.921

Table 5.5 Execution time values for a processor star using T4's
Case Study: 14 tasks, grain size (1 - 8).

T800-20	Number of processors per array			
Tasks Size	1	2	3	4
Size 8	2.0	1.097	0.858	0.751
Size 7	1.767	0.978	0.776	0.676
Size 6	1.537	0.859	0.686	0.606
Size 5	1.304	0.742	0.601	0.535
Size 4	1.074	0.631	0.517	**0.468**
Size 3	0.848	0.515	**0.426**	0.394
Size 2	0.609	**0.399**	0.342	0.323
Size 1	**0.382**	0.282	0.257	0.248

Table 5.6 Execution time values for a processor star using T8's
Case Study: 14 tasks, grain size (1 - 8).

The embolded values represent comparable times to execute the program on p processors for a problem needing k times more processing. Assuming a scaling factor $k = p$, and using Eq. (5.8),(5.9) and (5.10), the scaled metrics for this problems are calculated. These are summarised in Table 5.7.

System	p	Time (ms)	s_k	e_k	f_k	kf_k
Procstar-T4	1	1.917	-	-	-	-
Procstar-T4	2	1.930	1.987	0.994	0.007	0.014
Procstar-T4	3	2.057	2.796	0.932	0.037	0.111
Procstar-T4	4	2.234	3.432	0.858	0.055	0.220
Procstar-T8	1	0.382	-	-	-	-
Procstar-T8	2	0.399	1.915	0.958	0.044	0.088
Procstar-T8	3	0.426	2.690	0.897	0.058	0.174
Procstar-T8	4	0.468	3.265	0.816	0.075	0.300

Note: p=#processors, s_k=scaled speedup, e_k=scaled efficiency, f_k=scaled serial fraction

Table 5.7 Performance Summary of Processor Star Topology for a Scaled Problem (size 1-size 8) on T4's.

The scaled metrics generated data corresponding to speedup, that was closer to the linear case than that obtained using standard metrics. It was thus much easier to achieve efficient parallel performance than is implied by the conventional formulas. Table 5.7 shows the speedup data for the T4 implementation ranging from 1.987 to 3.432, at efficiencies from 0.994 to 0.858. Values slightly lower than these, were obtained for the T8 case. In both cases, there is a decrease in the efficiency e_k as p increases, and a smooth increase in f_k. This decreasing efficiency seems to be due to the fact that the problem size is not growing fast enough to completely counter the loss of efficiency as more processors are added. For the reasons presented here, the use of scaled metrics offer a suitable way to identify efficient speedup, not obtained by means of standard formulas.

5.6 SUMMARY

This Chapter has analysed the performance of dynamic and static task allocation tools, developed to automate the mapping of control systems on parallel processing transputer-based architectures.

A parallel version of a digital flight control law (VAP Control Law) has been mapped, using both dynamic and static tools, onto a parallel processing transputer-based systems, to give insight into the potential performance, improvements and bottlenecks associated with these strategies. In its basic realisation, for this particular controller example, process farming displayed an inferior performance when compared with the static approach.

Additional improvements to the original process farm were achieved implementing task buffering. This was efficient only for a small number of workers in the farm. With a larger number of workers the buffering version of the farm performed worse than the ordinary processor farm. An extended version, using the triple chain of workers, offered further improvement in performance, reducing significantly communication overheads and, thus, execution time. In this case the master was able to allocate tasks to each line and receive data results in parallel. The static task allocation approach, using a star topology, exhibited the best performance, having the advantage that there was no overhead associated with continually allocating the tasks in real-time.

Standard metrics were used for measuring the performance of the tasks allocation tools running on parallel processing transputer-based systems. Also a novel metric, the serial fraction, was utilised and provided information not revealed by other commonly used metrics.

In order to analyse and evaluate the performance of the tools in a more general way, since the results can vary strongly with the algorithm and size of the problem, this study has been further extended, considering more complex algorithms. For example, the effect of task granularity was studied. Through experiment in varying the size of task granularity in a controlled manner, it was confirmed that appropriate matching of tasks size to processor architecture results in efficient implementations.

Modern approaches for measuring parallel processor performance state that performance can be measured by scaling the problem to the number of processors, not by fixing the problem size, as it is normally used. When given a more powerful processing capacity, the problem generally should expand to make use of the increased facilities, keeping the execution time more or less constant. This modern approach was also considered for measuring the performance of the software tools developed.

CHAPTER 6

CONCLUDING REMARKS

6.1 GENERAL REVIEW

In this book we have attempted to provide an introduction to the application of parallel processing in real-time control. Following a general introduction to parallel processing, we have identified the Inmos transputer and its associated parallel programming language, occam, as, arguably, the most suitable, currently available, parallel hardware and software vehicles for embedded control systems. Inevitably, therefore, the book includes frequent, though not exclusive, reference to their use.

The issues concerning the effective use of parallel processing in control are too numerous to be satisfactorily covered here and we have, therefore, paid special attention to the following topics:

- A number of new strategies have been introduced for the implementation of concurrent control algorithms. This has been achieved by partitioning the algorithm into a number of tasks which may then be statically allocated onto a number of processing elements. The strategies include the Hybrid, the Parallel State-Space, and the FSVD approaches. The Hybrid strategy is a combination of two earlier implementations: the Heuristic and the Parallel Branches approaches. These strategies are assessed with respect to execution speed and ease of implementation.

- The popular control system design package, MATLAB, has been integrated with the Transputer Development System, TDS, to generate **EPICAS**, an Environment for Parallel Implementation of Control Algorithms and Simulation. This environment offers the control engineer a number of software tools for automating the implementation of control algorithms and simulation systems on transputer-based architectures.

Refer to the end of this Chapter for details of how to obtain EPICAS.

- A MATLAB toolbox has been described which automates the parallel partitioning of control algorithms into independent modules or tasks. The toolbox includes utilities for data entry, block partitioning and parallel representation.

- An OCCAM toolbox for automating the task allocation onto transputer-based systems has been presented. This toolbox includes utilities that implement either dynamic and static task allocation strategies on different topologies. These strategies can then be evaluated by means of displaying, on-line, task allocation data, processor activity and execution times.

- There has been a detailed performance analysis of using parallel processing in the implementation of a particular problem, the theme example - the fixed-size VAP control law. This has provided a clear insight into the potential performance improvements and bottlenecks of the different parallel processing structures used.

- This performance analysis has been extended to include the implementation of more complex models with a larger number of tasks, and varying granularity.

Besides computational speedup issues, topology choice may also be influenced by other concerns such as fault tolerance and hardware uniformity. Existing topologies, used to map systems onto a transputer array, suffer from the susceptibility of potential failure when one or more transputers fail. Fault-tolerance techniques, both in software and hardware, have been considered in [80]. In the event of a hardware failure, these methods enable each processor to re-organise and assume a greater portion of the processing task in the resulting system. An operational fault-tolerant flight controller based on the transputer star topology has been implemented. Multiple transputers have been used both to accommodate the parallelism of the algorithm and also to provide fault tolerance during in-service operation. This work is not reported here as the area of parallel processing in control and fault tolerance is covered in another book in this series [81].

6.2 FUTURE WORK

We hope that this book serves as a stimulus for further work and include some thoughts on possible improvements to parallel processing hardware and software and extensions of the mapping approaches and environment described here.

Occam and the Transputer

In general, the occam language and the transputer have exhibited high-performance potential in the implementation of real-time control systems. Occam has promoted structured modular programming and supported a sound paradigm for parallel processing and synchronisation, enhanced by the formal methods approach used to derive the language.

The CSP model has a number of advantages for the control engineer in the implementation of control systems. Control engineers commonly represent systems as diagrams of interconnected blocks. Occam offers an easy way of translating a block diagram description of a control system into a collection of concurrent processes.

The CSP model directly reflects the structure of these types of systems. However, although occam can easily implement control systems structures, it does not offer implicit processor task allocation. The present implementation of occam forces the programmer to explicitly perform the mapping of processes to processors, and communication channels to hardware links, by means of configuration statements. Future occam implementations should provide automatic process scheduling without requiring explicit processor assignment from the programmer.

The transputer has been developed with dedicated hardware attributes to support both parallel processing and the occam model of CSP. Thus the compilation of occam into transputer executable code is very efficient. However, as the transputer is a general-purpose microprocessor architecture, it is suitable for problems involving coarse to medium task granularity in order to minimise communication overheads. This is suitable for computationally complex control algorithms. On the other hand, for simple control applications, involving short time scales and reduced numbers of operations, parallel processing using transputers is not suitable because communication overheads will dominate the total execution time. Here, a finer-grain approach to parallel processing may be more suitable [82].

Developments arising out the EEC Open Microsystems Initiative might prove significant here since users may be able to construct custom parallel processors of appropriate granularity to match their application.

In this context, we must also remember that, as well as task granularity and processor granularity, "communications granularity" [83] is also an important consideration. For example, inter-transputer communication is more efficient when longer messages are sent infrequently rather than a large number of small messages. Communications granularity is a measure of how communications speed and the handling of message size varies from processor to processor.

MATLAB Toolbox extensions

In the development of MATLAB software tools within EPICAS, further work can be realised in the partitioning, discretisation and parallel representation processes. Here, the Parallel State-Space approach has been used as the most appropriate strategy for the partitioning of a control algorithm. Alternatively, other strategies such as the Heuristic, Parallel Branches, Hybrid, and FSVD methods can be implemented. These strategies may be more appropriate for other computer architectures. The use of a Gantt chart, for example, in the support of the Heuristic strategy has already been reported in [84] for Kalman filtering and self-tuning applications.

The Pole-Zero Mapping method and its variants have been used for the discretisation of the analog models. However, other discretisation methods (e.g. the Tustin transformation, z-transform methods, etc.) might prove more suitable in their handling of specific transient response and frequency response characteristics. Alternatively, optimisation methods can be used to minimize some performance error criteria between the continuous and discrete system. (The use of optimisation methods has the benefit that the overall system characteristics can be considered in the process instead of considering the controller as a stand-alone system.)

OCCAM Toolbox Developments

We have seen that our representation of the control algorithm as a number of independent first- and second-order equations, and their subsequent utilisation as the primary grain size of the parallel tasks, may ultimately prove too fine a grain description. Schemes should be developed to support the generation of tasks of higher order in order to find the best grain size to match the requirements of performance for a given system.

In the development of OCCAM software tools for parallel implementation on transputer-based systems, one of the problems we have faced is the issue of portability which results from the fact that transputer systems require a host front end, and there are different host computers available (IBM PC, SUN, ATW, Apollo, Macintosh and other workstations). This means that there is yet another dimension to the hardware platform.

The tools to automate the mapping of control systems on transputer systems have been originally developed on a PC platform, trying to avoid host-dependent user interfaces because these interfaces are not portable. However, the basic input/output procedures of the OCCAM mapping tools are host-dependent. Currently, work is proceeding to develop a more portable system based on the Occam 2 Toolset. This work could also be continued to accommodate the use of languages, other than occam, in the implementation of a controller.

Real-Time Simulation

In the field of real-time simulation, the environment could be adapted to allow the control engineer to evaluate the on-line performance of a control algorithm implemented in a parallel processing system by embedding this controller in a simulation of the total system where the controller will be located. Further, we have already shown, in Chapter 4, that this simulation itself may also be implemented on a multiple transputer network.

Applications Software Developments

Of course, the calculation of the control law forms only a component, albeit a critical one, of the real-time control software function. It is envisaged that this environment will form part of a larger scheme involving the mapping of general control software onto a parallel processing system, which may also include fault-tolerant features. Manson [85] is developing an approach, using the CASE tool, Software through Pictures (StP), as a harness, to represent occam processes in data flow diagram form, and, ultimately, to generate occam code, configured to execute on the desired topology. Croll [86] is using StP in his work on the development of safety-critical software. Clearly, the mapping procedures and performance analysis techniques described here could be integrated into such schemes.

6.3 SUMMARY

Parallel processing offers a suitable means of achieving the computing capability required in the implementation of real-time control systems, but only if an appropriate grain-size and a correct scheduling strategy for the target parallel processing architecture are utilised. Additionally, its inherent fault-tolerance characteristics makes parallel processing a natural way to construct ultra-reliable systems.

While it is our view that parallel processing exhibits great promise for the satisfaction of the growing demands of control system designers, there are many exciting opportunities for control engineers to develop this potential to its full extent.

For distribution details, concerning the EPICAS software, contact authors:

Dr. Fabian Garcia Nocetti
SEES, University of Wales, Bangor
Dean Street, Bangor, Gwynedd,
LL57 1UT, U.K.
Tel. +44 (0)248 351151
Fax. +44 (0)248 361429

Prof. Peter Fleming
Automatic Control and Systems Engineering,
University of Sheffield,
PO BOX 600, Mappin Street, Sheffield
S1 4DU, U.K.
Tel: +44 (0)742 768555
Fax: +44 (0)742 731729

REFERENCES

1. Speyer, J.L., "A Perspective on Aerospace Control Systems", IEEE Control Systems Magazine, April 1987, pp 11-13.

2. McLean, D., "Active Control of a Modern Fighter Aircraft", Science and Engineering Research Council (SERC) Vacation School, Computer Control, University of Sheffield, March 1987, pp. 197-214.

3. Travassos, R. and Kaufman, H., "Impact of Parallel Computers on Adaptive Flight Control", Proceedings of JACC, San Francisco CA, USA, 1980, pp. WP1-B.

4. Jaswa, V.C. and Thomas, C.E., "CPAC -Concurrent Processor Architecture for Control", IEEE Transactions on Computers, Vol. C-34, No. 2, February 1985, pp. 163-169.

5. Fleming, P. J., ed. "Parallel Processing in Control -the Transputer and other Architectures", Peter Peregrinus Ltd, 1988.

6. Garcia Nocetti, D.F., "Parallel Implementation of a Flight Control Law", M.Sc. Thesis, University of Wales, Bangor, U.K., 1987.

7. Roberts, R.A. and Mullis, C.T., "Digital Signal Processing", Addison-Wesley, 1987.

8. Moler, C., Little, J., Bangert, S. and Kleinman, S., PC-MATLAB User's guide, The MathWorks Inc. 1987.

9. Goddard, K.F., "Theoretical Studies of Automatic Control Laws for a BAC 1-11 Aircraft Utilising the Wing Spoilers for Direct Lift Control", Technical Report 79034, Royal Aircraft Establishment, 1979.

10. Goddard, K.F. and Cooke, N. "Flight Trials of an Automatic Control Law for a BAC 1-11 Aircraft", Technical Report 80003, Royal Aircraft Establishment, 1980.

11. Manson, G.A., "The Granularity of Parallelism Required in Multi-Processor Systems", Science and Engineering Research Council (SERC) Vacation School, Computer Control, University of Sheffield, March 1987, pp. 96-111.

12. Flynn, M.F., "Some Computer Organisations and Their Effectiveness", IEEE Trans. Comput., C-21, 1972, pp. 948-960.

13. Stone, H.S., "High Performance Computer Architectures", Addison Wesley, 1987.

14. Seitz, C.L., "The Cosmic Cube", Communications of the ACM, Vol.28, No.1, 1985, pp.22-33.

15. Pfister, G.F., Brantley, W.C., et. al., "The IBM Research Parallel Processor Prototype (RP3):Introduction and Architecture", Proc. International Conference on Parallel Processing, August 1985, pp. 764-711.

16. Dettmer, R., "Chip Architectures for Parallel Processing", Electronics and Power, Vol. 31, No. 3, April 1985, pp. 227-231.

17. Burns, A., "Concurrent Programming in Ada", Ada Companion Series. Cambridge University Press, 1985.

18. INMOS Limited, "Transputer Overview", The Transputer Databook, Second edition, September 1989.

19. Harp, G. (ed), "Transputer Applications", Pitman, 1989.

20. Fleming, P.J. (ed), "Parallel Processing in Control - the Transputer and other Architectures", Peter Peregrinus, 1988.

21. Dyson C., "The Next Generation Transputer", Technical Note, Inmos Bristol, July 1990.

22. Inmos Ltd, "Occam 2 Reference Manual", Prentice Hall, 1987.

23. Bakkers, A.W.P. and Van Amerongen, J. "Transputer-based Control of Mechatronic Systems", Proc. 11th IFAC World Congress, Tallinn, USSR, 1990.

24. Thielemans, H. and Verhulst, E, "Implementation Issues of Trans-RTXc on the Transputer, Proc. IFAC Workshop on Algorithms and Architectures for Real-Time Control, Bangor, UK, Pergamon Press, 1991.

25. Welch, P.H. "Multi-Priority Schedulers for Transputer-based Real-time Control", Real-Time Systems with Transputers, IOS Press, 1990.

26. Mirab, H. and Gawthrop, P.J., "Transputers for Robot Control, 2nd International Transputer Conference, Antwerp, Bira, 1989.

27. Jones, D.I. and Entwistle, P.M., "Parallel Computation of an Algorithm in Robotic Control", Proc IEE Conf. Control 88, University of Oxford, Oxford, U.K., April, 1988, pp. 438-443.

28. Da Fonseca, P., Entwistle, P.M. and Jones, D.I., "A Transputer Based Processor Farm for Real-time Control Applications", Proc 2nd Int. Conf. on Applications of Transputers, Southampton, UK, 1990, pp. 140-147.

29. Daniel, R.W. and Sharkey, P.M., "The Transputer Control of a Puma 560 Robot via the Virtual Bus", Proc. IEE, Part D, 1990, vol 137, pp 245-252.

30. Jones, D.I. and Fleming P.J., "Control applications of transputers", Chapter 7 in Parallel Processing in Control - the Transputer and other Architectures, Peter Peregrinus, 1988.

31. Asher, G.M. and Sumner, M., "Parallelism and the Transputer for Real Time High Performance Control of AC Induction Motors", Proc. IEE, Part D, 1990, vol 137, pp 179-188.

32. Entwistle, P. M., "Parallel Processing for Real-Time Control", PhD Thesis, University of Wales, Bangor, U.K., 1990.

33. Maguire, L.P., "Parallel Architectures for Kalman Filtering and Self-tuning Control", PhD Thesis, The Queen's University of Belfast, 1991.

34. Lawrie, D.I., Fleming P.J., Irwin, G.W. and Jones, S., "Kalman Filtering: a Survey of Parallel Processing Alternatives", Proc IFAC Workshop on Algorithms and Architectures for Real-Time Control, Bangor, UK, 1991.

35. Bahramparvar, M.R. and Gray, J.O. "Application of Parallel Processing Techniques to Eddy-Current NDT Instrumentation", Proc. IEE Part D, vol 137, 1990, pp. 211-224.

36. Atherton D.P., Gul, E., Kountzeris, A. and Kharbouch, M., "Tracking Multiple Targets Using Parallel Processing", Proc. IEE Part D, Vol 137, 1990, pp. 225-234.

37. Garcia Nocetti, D.F., Thompson, H.A., De Oliveira, C.M. Jones and P.J. Fleming, "Implementation of a transputer based flight controller", IEE Proc., Part. D, Vol. 137, 1990, pp. 130-136.

38. Thompson, H.A. and Fleming, P.J., "Fault Tolerant Transputer-Based Controller Configurations for Gas Turbine Engines", Proc. IEE, Part D, Vol. 137, 1990, pp. 253-260.

39. Rogers, E. (ed) Special Issue on "Parallel processing in control", Int. J. Control, Vol. 54, No. 6. 1991.

40. Irwin, G.W. and Fleming P.J. (eds), "Transputers for Control" Research Studies Press, John Wiley & Sons Ltd, 1992.

41. Sarkar, V., "Partitioning and Scheduling Parallel Programs for Multiprocessors", Pitman Publishing, London, 1989.

42. Cvetanovic, Z., "The Effects of Problem Partitioning, Allocation, and Granularity on the Performance of Multiple-Processor Systems", IEEE Transactions on Computers, Vol. C-36, No. 4, April 1987, pp. 421-432.

43. Shaffer, P., "Experience with Implementation of a Turbojet Engine Control Program on a Multiprocessor", Proc. American Control Conference, 1989, pp. 2715-2720.

44. Kasahara, H., and Narita, S., "Practical Multiprocessor Scheduling Algorithms for Efficient Parallel Processing", IEEE Transactions on Computers, Vol. C-33, No. 11, November 1984, pp. 1023-1029.

45. Lee, S.Y. and Aggarwal, J.K., "A Mapping Strategy for Parallel Processing", IEEE Transactions on Computers, Vol. C-36, No.4, April 1987, pp. 433-441.

46. Berger, M. and Bokhari, S., "A Partitioning Strategy for Nonuniform Problems on Multiprocessors", IEEE Transactions on Computers, Vol. C-36, No. 5, May 1987, pp. 570-580.

47. Bokhari, S.H., "Assignment Problems in Parallel and Distributed Computing", Kluwer Academic Publishers, 1987.

48. Fleming, P.J., Garcia Nocetti, D.F. and Thompson, H.A., "Implementation of a Transputer-based Flight Controller". Proc. IEE Conf. Control 88, University of Oxford. Oxford, U.K. April 1988.

49. De Oliveira, M.C.F., "CAD Tools for Digital Control", PhD Thesis, University of Wales, Bangor, 1990.

50. Franklin, G.F. and Powell, J.D., "Digital Control of Dynamic Systems", Addison-Wesley, 1981.

51. Franklin, G.F. and Powell, J.D. "Digital Control of Dynamic Systems", Addison-Wesley, Second Edition, 1989.

52. Moler, C., "MATLAB User's Guide", Department of Computer Science, University of New Mexico, Albuquerque, USA, 1980.

53. Moler, C., Little, J. and Bangert, S., "PRO-MATLAB User's Guide", MathWorks, Inc., August 1987.

54. Mathworks Newsletter, The MathWorks Inc., Volume 3, No. 2, May, 1989.

55. Laub, A., Little, J., "Control Systems Toolbox for use with MATLAB, User's Guide", The MathWorks, Inc., August, 1986.

56. Chiang, R., Safonov, M., "Robust-Control Toolbox for use with MATLAB, User's Guide", The MathWorks, Inc., June 1988.

57. Boyle, J.M., Ford, M.P. and Maciejowski, J.M., "Multivariable Frequency Domain Toolbox for use with MATLAB, User's Guide", GEC Eng. Research Centre and Cambridge Control LTD. April, 1988.

58. Little, J. and Shure, L., "Signal Processing Toolbox for use with MATLAB, User's Guide", The MathWorks Inc., August, 1988.

59. Ljung, L., "System Identification Toolbox for use with MATLAB, User's Guide", The MathWorks Inc., April 1988.

60. Milne, G., "State-Space Identification Toolbox for use with MATLAB, User's Guide", The MathWorks Inc., March 1988.

61. Ogata, K., "Discrete-Time Control Systems", Prentice-Hall International Inc., 1987.

62. Franklin, G.F., Powell, J.D. and Workman, M.L., "Digital Control of Dynamic Systems Toolbox for use with MATLAB", The Mathworks Inc., December 1989.

63. Simulab, User's Guide, The MathWorks, Inc., September 1990.

64. Protoblock User's Guide, Grumman Aerospace Coorporation, October 1989.

65. Entwistle, P., Lawrie, D.I., Thompson, H.A., and Jones, D.I., "An Eurocard Computer using Transputers for Control Systems Applications". IEE Colloquium on Eurocard computers -a solution to low-cost control', September, 1989.

66. Inmos Limited, "IMS B004 Evaluation Board", User Manual, November 1985.

67. Inmos Transputer Development System IMS D700D, User Manual, Prentice Hall, 1988.

68. Inmos Limited, "Transputer Development Systems and iq Systems Databook", First Edition 1989.

69. Hill, G., "Designs and Applications for the IMS C004", INMOS Technical Note 19, June 1987.

70. Jones, C.M., "Digital Flight Controller Implementation using Parallel Processing Techniques", Final Report of Agreement 2025/26/RAE(B), University of Wales, Bangor, August 1990, pp. 24-34.

71. TSIM, Nonlinear Dynamic Simulation Package User Manual, Cambridge Control LTD, Cambridge, England, 1989.

72. Parkinson, D. and Liddell, H.M., "The Measurement of Performance on a Highly Parallel System", IEEE Transactions on Computers, Vol. C-32, No. 1, January 1983.

73. Flatt, H.P. and Kennedy, K., "Performance of Parallel Processors", Parallel Computing 12, October 1989, pp. 1-20.

74. Graham, J.H. and Kadela, T.F., "Parallel Algorithms and Architectures for Optimal State Estimation. IEEE Transactions on Computers, Vol. C-34, No. 11, November 1985.

75. Karp, A.H. and Flatt, H.P., "Measuring Parallel Processor Performance", Communications of the ACM, Vol 33, 5, May 1990, 539-543.

76. Gustafson, J.L., "Reevaluating Amdahl's Law", Communications of the ACM 31, 5, May 1988, pp. 532-533.

77. Karp, A.H. and Flatt, H.P., "Measuring Parallel Processor Performance", Communications of the ACM, Vol 33, 5, May 1990, 539-543.

78. Gustafson, J.L., "Reevaluating Amdahl's Law", Communications of the ACM 31, 5, May 1988, pp. 532-533.

79. Green, S.A. and Paddon, D.J., "An Extension of the Processor Farm using a Tree Architecture", Proc. of the 9th OUG Technical Meeting, Southampton, U.K., September 1988.

80. Garcia Nocetti, D.F., "Parallel Processing in Digital Flight Control", PhD Thesis, University of Wales, Bangor, U.K., 1991.

81. Thompson, H.A., "Parallel Processing for Jet Engine Control", Advances in Industrial Control Series, Springer-Verlag London Ltd, 1992.

82. Xu, Y., Jones, S., Spray, A., et al. "Implementing Control Algorithms in Pace", submitted to Proceedings IEE, Part D, Control Theory and Applications, January, 1992.

83. Dodds, G.I., Private communication, Queen's University of Belfast, 1991.

84. Maguire, L.P., "Parallel Architectures for Kalman Filtering and Self-tuning Control, PhD dissertation, The Queen's University of Belfast, 1991.

85. Manson, G.A., "Towards a CASE Tool for Parallel Systems", Colloqium on Computer Aided Software Enginnering Tools for Real-time Control, IEE Digest No. 1991/087, April 1991, pp 7/1-7/4.

86. Croll, P.R., Private Communication, Sheffield University, 1992.

INDEX

bus-contention 12
communications granularity 135
concurrent languages 14
CSP model 134
discretisation 44
EPICAS 27, 51, 133, 138
folding editor 21, 64
Flynn's taxonomy 9
granularity issues 111   117
    higher granularity models 117
    scaled-sized models 126   112
    variable number of tasks 112
M-files 53
mapping strategies 33, 44
    FSVD 42, 132
    Heuristic 35, 132
    Hybrid 39, 132
    Parallel Branches 37, 133
    Parallel State-Space 132
    performance 47
    transputer implementation 45 45
MATLAB 53
MIMD architectures 10
    loosely-bound 11
    tightly-bound 11
Occam language 13, 18
    channels 19

constructs 20
prioritisation 23
processes 19, 21
Occam programming system (OPS) 21
parallel architectures 8
    MIMD 10
    MISD 10
    SIMD 9
    SISD 8
parallel processing 6, 8
parallel programming 7, 18
partitioning 30, 33, 53, 58, 89
PEs (processing elements) 6
process mapping 14, 24
processor farm model 67, 73
    buffered approach 101
    extended topology 103
    linear topology 67, 97
    processor activity 74, 106
    transputer implementation 68, 98, 102, 107
    performance 96
processor star approach 76
    processor activity 80
    transputer implementation 77, 110
    performance 108
processor-memory configurations 12
Protoblock 61

R/C ratio 11, 92
real-time kernels 24
real-time programming 23
scaled metrics 95, 129
sequential programming 18, 19
Simulab 61
Simulation 83
    aircraft flight dynamics 84, 87
    BAC 1-11 equations of motion 84, 85
    model response 86, 88
    using EPICAS 87
    using TSIM 85
standard metrics 92
    efficiency 47, 94
    execution time 47, 93
    price/performance 92, 93
    serial fraction 76, 94
    speedup 47, 93
task allocation 24, 30, 63
task granularity 11, 92
    coarse-grain 11, 92
    fine-grain 11, 92
    medium-grain 11

tools for parallel partitioning 54
    data entry process 55
    parallel tasks representation 60
    partitioning process 58
tools for task allocation 63
    dynamic task allocation 66
    MATLAB-OCCAM interface 66
    static task allocation 76
transputer 12, 14
    architecture 16
    concurrency 14, 16
    T800 version 15
    T9000 series 18
transputer applications 25
    fault-tolerance 28
    flight control 27, 30
    Kalman filtering 27
    motor control 26
    robotics 25
Transputer Development System (TDS) 64
transputer topologies 17, 63
VAP control law 31, 55
von Neumann bottleneck 12